USMLE Step I

BrainChip for Biochemistry

Notice: The indications and dosages of all drugs in this book have been recommended in the medical literature and conform to the practices of the general community. The medications described and treatment prescriptions suggested do not necessarily have specific approval by the Food and Drug Administration for use in the diseases and dosages for which they are recommended. The package insert for each drug should be consulted for use and dosage as approved by the FDA. Because standards for usage change, it is advisable to keep abreast of revised recommendations, particularly those concerning new drugs.

USMLE Step 1 Review

BrainChip for Biochemistry

Scott Lee
UCLA School of Medicine
Los Angeles, California

Michael Turner, Lead author
Harvard University School of Medicine
Boston, Massachusetts

Grant Lee
Stanford University
Stanford, California

Blackwell
Publishing

© 2002 by Blackwell Science, Inc.
a Blackwell Publishing Company

EDITORIAL OFFICES:
Commerce Place, 350 Main Street, Malden, Massachusetts 02148, USA
Osney Mead, Oxford OX2 0EL, England
25 John Street, London WC1N 2BS, England
23 Ainslie Place, Edinburgh EH3 6AJ, Scotland
54 University Street, Carlton, Victoria 3053, Australia

OTHER EDITORIAL OFFICES:
Blackwell Wissenschafts-Verlag GmbH, Kurfürstendamm 57, 10707 Berlin, Germany
Blackwell Science KK, MG Kodenmacho Building, 7-10 Kodenmacho Nihombashi, Chuo-ku, Tokyo 104, Japan
Iowa State University Press, A Blackwell Science Company, 2121 S. State Avenue, Ames, Iowa 50014-8300, USA

DISTRIBUTORS:

The Americas
   Blackwell Publishing
   c/o AIDC
   P.O. Box 20
   50 Winter Sport Lane
   Williston, VT 05495-0020
   Telephone orders: (800) 216-2522
   Fax orders: (802) 864-7626
Australia
   Blackwell Science Pty. Ltd.
   54 University Street
   Carlton, Victoria 3053
   Telephone orders: 03-9347-0300
   Fax orders: 03-9349-3016

Outside The Americas and Australia
   Blackwell Science, Ltd.
   c/o Marston Book Services, Ltd.
   P.O. Box 269
   Abingdon
   Oxon OX14 4YN
   England
   Telephone orders: 44-01235-465500
   Fax orders: 44-01235-465555

All rights reserved. No part of this book may be reproduced in any form or by any electronic or mechanical means, including information storage and retrieval systems, without permission in writing from the publisher, except by a reviewer who may quote brief passages in a review.

Acquisitions: Beverly Copland
Development: Julia Casson
Production: GraphCom Corporation
Manufacturing: Lisa Flanagan
Marketing Manager: Kathleen Mulchay
Printed and bound by Sheridan Books

Printed in the United States of America
02 03 04 5 4 3 2 1

The Blackwell Science logo is a trade mark of Blackwell Science Ltd., registered at the United Kingdom Trade Marks Registry.

Library of Congress Cataloging-in-Publication Data

Lee, Scott, 1974-
  BrainChip for biochemistry / by Scott Lee, Michael Turner, Grant Lee.
    p. ; cm.
  ISBN 0-632-04636-8 (pbk.)
    1. Biochemistry—Outlines, syllabi, etc. 2. Clinical biochemistry—Outlines, syllabi, etc.
3. Physicians—Licenses—United States—Examinations—Study guides.
    [DNLM: 1. Biochemistry—Outlines. QU 18.2 L481b 2002] I. Title: Brain chip for biochemistry. II. Turner, Michael, 1975- III. Lee, Grant. IV. Title.
  QP514.2 .L42 2002
  572—dc21

20010068

Table of Contents

Preface, *xi*
Dedication and Acknowledgments, *xiii*

1 General Principles of Biochemistry, *1*
a. An enzyme can be either "stimulated" or "inhibited," *1*
b. An enzyme can be "activated" or "deactivated" by another enzyme, *1*
c. Receptors can be "up-regulated" or "down-regulated," *1*
d. Agents in a pathway can be "upstream" or "downstream," *1*
e. Rate-limiting step of a reaction pathway—the slowest step, *1*
f. Oxidation versus reduction, *2*
g. Kinase versus phosphatase enzymes, *2*
h. Both formation <u>and</u> degradation of cAMP happen at the same time, *2*

2 Acid-Base FUNdamental Concepts, *3*
a. Equilibrium, *3*
b. Acids and their conjugate bases, *3*
c. pH, *4*
d. Buffers, *4*
e. The physiologic buffer system, *6*

3 DNA and RNA—Structure and Important Processes, *8*
a. DNA: What is it? Who cares? *8*
b. Nitrogenous bases (N bases): What you need to know, *8*
c. Nucleotides: Basic structural unit of DNA and RNA: A molecule with three important parts, *8*
d. Structure of DNA, *9*
e. Structure of RNA, *12*
f. Structure of a gene, *12*
g. Overview: Key ideas of replication, transcription, and translation, *13*
h. DNA replication, *13*
i. Transcription, *16*
j. Three different structures made from RNA, *18*
k. Regulation of transcription, *19*
l. Translation, *20*
m. Summary and comparison of replication, transcription, and translation, *22*
n. Types of DNA mutations and how they affect protein structure, *22*
o. Clinical correlations, *24*

4 Techniques of Biotechnology and Molecular Biology, *26*
a. Procedures involving DNA and/or RNA, *26*
b. You need to understand four procedures involving proteins, *27*
c. Procedures involving proteins, *28*

5 Proteins, 31

  a. Protein structure: What you need to know, 31
  b. The four levels of protein structure—an overview, 32
  c. The four levels of protein structure—detailed explanations of important points, 35
  d. Protein packaging and distribution, 36
  e. Collagen: The most abundant protein in the body, 37
  f. Porphyrins: Cyclic polypeptide compounds that bind metal ions, 38
  g. Heme: A porphyrin that binds iron, 41
  h. Quick clinical facts, 43
  i. Myoglobin: A reservoir of oxygen for cardiac and skeletal muscle, 44
  j. Hemoglobin: A protein that binds and transports oxygen in the blood, 45
  k. Oxidized forms of hemoglobin and myoglobin, 48

6 Enzymes, 49

  a. How do cells capture and store energy? 49
  b. ΔG: The free-energy change involved in a certain reaction, 49
  c. The free-energy prediction game, 53
  d. Enzymes: Proteins that increase the rate of a reaction by lowering the activation energy, 53
  e. Michaelis-Menten equation, 55
  f. Allosteric enzymes have multiple active sites that interact with each other, 56
  g. What physiological strategies does the body use to increase the rate of a reaction? 58
  h. How can the function of a single enzyme be inhibited? 59
  i. Competitive versus noncompetitive inhibition, 60
  j. Increase in enzyme amount, 61

7 Cellular Second-Messenger Systems, 62

  a. Interactions between extracellular molecules and the intracellular environment, 62
  b. Overview: The five most physiologically important second-messenger systems, 63
  c. Detailed summary: The five most physiologically important second-messenger systems, 64
  d. Summary: Comparing and contrasting the three types of G protein-linked second-messenger cascades, 64
  e. Malfunctions of G protein–linked second-messenger systems, 64

8 Overview of Metabolism With Emphasis on Harmonal Control, 69

  a. General principles, 69
  b. Catabolism: Breaking down substrates for energy, 73
  c. Anabolism: Synthesizing important materials, 78
  d. Three major metabolic states and the main reactions occurring in different tissues, 78
  e. Citrate: An important metabolic player, 82
  f. Pyruvate: A crucial substrate at a metabolic crossroads, 82
  g. Lactate, 82
  h. Acetyl CoA: Another crucial substrate at a metabolic crossroads, 83
  i. Glucose metabolism: Up close and personal, 83
  j. Insulin: The body's means of storing fuel substrates in response to the "well-fed" state, 85
  k. Glucagon: Releases fuels into the blood and sequesters amino acids in the liver for gluconeogenesis, 89

l. Summary: Insulin and glucagon have opposing effects on the blood levels of three important substrates, *92*
m. The liver: It does everything but iron your socks, *92*
n. The brain: A metabolically unique tissue, *93*

9 TCA CYCLE AND ELECTRON TRANSPORT CHAIN, *96*

a. The TCA cycle: Generating NADH and $FADH_2$ to send to the electron transport chain, *96*
b. The electron transport chain: Turning NADH and $FADH_2$ into ATP, *101*

10 CARBOHYDRATE METABOLISM, *111*

a. An overview of glycogen synthesis and regulation, *111*
b. Glycogen storage diseases, *113*

11 LIPID AND CHOLESTEROL METABOLISM, *139*

a. General principles, *139*
b. All about fatty acids, *139*
c. De novo FFA synthesis: An anabolic pathway that takes acetyl CoA and turns it into palmitate, *140*
d. Beta oxidation: A catabolic pathway that takes an FFA and repeatedly removes two carbon fragments to make acetyl CoA, NADH, and $FADH_2$, *142*
e. Summary and comparison of de novo FFA synthesis versus beta oxidation, *145*
f. Key principles of TG storage and metabolism, *146*
g. TG synthesis: An anabolic pathway that converts three FFAs and one glycerol-phosphate into a triglyceride, *146*
h. TG storage: Takes newly synthesized TGs and stores them in adipose cells, *147*
i. TG degradation: Breaking down TGs to yield three FFAs and one glycerol, *147*
j. Exogenous and endogenous TG transport, *148*
k. Phospholipids: Two fatty acids and one polar group attached to a backbone of glycerol or sphingosine, *151*
l. Everything you need to know about bile, *152*
m. Everything you need to know about bilirubin, but were afraid to ask, *153*
n. Cholesterol: A crucial molecule that is made from acetyl CoA, *157*
o. Exogenous and endogenous circulation of cholesterol, *158*

12 PROTEIN AND NITROGEN METABOLISM, *167*

a. General principles, *167*
b. Ways to metabolically classify amino acids: (1) glucogenic, ketogenic, or both and (2) essential or nonessential, *169*
c. Transamination: Transfer of NH_3 between an amino acid and alpha-KG, *171*
d. Oxidative deamination: Removal of NH_3 from glutamate to yield alpha-KG, *173*
e. The glutamate/glutamine cycle: A way for tissues to send NH_{4+} to the liver or kidneys, *174*
f. The glucose/alanine cycle: A way for muscle cells to get rid of NH_3, and for liver cells to supply muscles with fresh glucose during periods of strenuous exercise, *175*
g. Review of ammonia (NH_{4+}) metabolism: It is a deadly toxin that can cross the BBB and harm the CNS. It must NOT be allowed to accumulate in the blood, *176*
h. Transamination and oxidative deamination WORK TOGETHER in both degradation and synthesis to maintain amino acid homeostasis, *177*
i. Urea cycle: The liver's way of converting ammonia into something nontoxic and excretable—urea, *179*

13 NUCLEOTIDE METABOLISM, *181*
a. General principles, *181*
b. The HMP shunt: A way for cells to divert glucose-6-P from glycolysis to make NADPH and ribose-5-phosphate, *182*
c. PURINES—de novo purine synthesis: Building a nucleotide from scratch, *184*
d. Purine synthesis by the salvage pathway: Take free N bases from the breakdown of RNA and DNA and add a sugar and a phosphate to them, *186*
e. Degradation of purine nucleotides: Strip the ribose and the phosphate groups away, then convert the N base to uric acid, *186*
f. PYRIMIDINES: De novo pyrimidine synthesis: Build a nucleotide from scratch, *187*
g. Pyrimidine synthesis by the salvage pathway: Take free N bases from the breakdown of RNA and DNA and add a sugar and a phosphate to them, *188*
h. Degradation of pyrimidines: Strip the ribose and the phosphate groups away, then convert the N base to acetyl CoA or succinyl CoA, *189*

14 INTEGRATED REVIEW OF METABOLISM, *189*
a. Intracellular locations of important metabolic processes, *189*
b. NAD^+ and NADPH: Two crucial substances needed for many metabolic reactions, *189*
c. Citrate: An important metabolic player. There are three important metabolic roles of citrate, *194*
d. Pyruvate: A crucial substrate at a metabolic crossroads, *194*
e. Lactate: What you need to know, *195*
f. acetyl CoA: Another crucial substrate at a metabolic crossroads, *196*
g. The enzyme name game: I name the enzyme; you name the metabolic process and reaction to which it corresponds, *198*
h. Intracellular shuttles: Crucial for moving materials between the cytosol and the mitochondrial matrix, *199*
i. Metabolic cycles: Crucial for allowing the liver to supply peripheral tissues with fuels, and for allowing these tissues to export $NH4^1$ to the liver, *201*
j. Metabolic consequences of not ingesting enough carbohydrates, *202*
k. Nutritional connection: There are certain biochemical rationales that support the idea of a low-carbohydrate diet, *202*
l. Nutritional connection: There are certain biochemical rationales against the idea of a low-carbohydrate diet, *202*

15 NUTRITION, *204*
a. High-yield concepts in nutrition, *204*
b. What is an RDA? *205*
c. The name game: Common names and scientific names of important vitamins, *206*
d. Important vitamins and their biologically active forms, *206*
e. What are the unique characteristics of water-soluble vitamins? *206*
f. What are the unique characteristics of fat-soluble vitamins? *207*
g. Important nutrients: Their functions and mechanisms, *207*
h. Everyone's favorite game: Name that deficiency, *210*
i. Quick reference summary, *210*

INDEX, *213*

PREFACE

BrainChip for Biochemistry, a continuation of the BrainChip series, was written out of the need for a complete review book in biochemistry, one that we could not find while studying for the boards. It is written in a relatively colloquial manner and in a way that we believe is the easiest for the reader to assimilate difficult concepts.

There are three major components to a high-yield book. First, a concise, succinct, and high-yield text that is complete—in and of itself—for USMLE Step 1 is a must in any review text. Second, such a text would cover the essential basic science of biochemistry with all its mechanistic details, pathways, and theoretical concepts. Third, it would also include clinical presentations of pathologies and their biochemical bases. We have attempted to include these three components in the text and have also provided illustrations as visual aids to the material at hand.

With an understanding that the student has a standard biochemistry course, *BrainChip for Biochemistry* is written for rapid review of biochemistry in 4 to 7 days with the most essential information provided. It is intended to be used in conjunction with a more comprehensive text if used for a course. With a sufficient background, however, it can be read alone.

BrainChip for Biochemistry is organized into fifteen chapters, each treating a major topic within biochemistry. A "history of the present illness" is included in each chapter, as it is often presented on examination. Only the most salient facts are included, which is often a very brief HPI. This approach may not be indicative of many boards' questions in which extraneous material is often presented.

Each HPI contains its most conspicuous information in bold lettering and is followed by the name of the disease and the biochemical mechanisms that are responsible. We have included what we believe is sufficient for Step 1, but examiners have a way of asking the most obscure facts. We advise that you use this text in conjunction with your class biochemistry notes, if you desire additional detail on any subject.

BrainChip for Biochemistry has been prepared with a great deal of research and corroboration with multiple authorities. However, as with any text, there are often shortcomings. We welcome and encourage your suggestions, advice, and vitriolic attacks on our corny humor (mainly Michael's).

It is our hope that you find this text beneficial and an efficient use of your time. We sincerely wish you success on Step 1 and your examinations.

Dedication and Acknowledgments

To the staff at Blackwell—Julia Casson, William Deluise, and Beverly Copland for their continuing support, dedication, and hard work.

To our mentors—Armand Fulco, Jonathan Kaunitz, Hynda Kleinman, Edward H. Livingston, Arun Patel, Deborah Philp, Joseph Pucci, and Ravi Tolwani for teaching us many things both inside and outside the classroom.

To our friends—Michael Burrow, Ugo Iroku, Jon Kim, Daniel Ko, Dinh Le, Josh Lee, Kristina Steenson, Troy Ward, and Chris Yoo, for their critical reading of the manuscript and our character, as well as their kindness and support, for without which this book would not have been published.

To our families—Melodie Mallory-Solton, who has faithfully lived out the meaning of unconditional love. Dong and Kyung Lee, for their love and unceasing devotion.

And finally—Soli deo gloria.

Abbreviations

Abx	Antibiotics
ADP	Adenosine diphosphate
AIDS	Acquired immunodeficiency syndrome
ATP	Adenosine triphosphate
BBB	Blood-brain barrier
BSE	Bovine spongioencephalitis
BUN	Blood urea nitrogen
CMV	Cytomegalovirus
CNS	Central nervous system
CPE	Cytopathogenic effect
C$_R$	Creatinine
CSF	Cerebrospinal fluid
CXR	Chest x-ray
CT	Computed tomography
DIC	Disseminated intravascular coagulation
DNA	Deoxyribonucleic acid
DTH	Delayed-type hypersensitivity
EBV	Epstein-Barr virus
EEE	Eastern equine encephalitis
EMB	Eosin methylene blue
FTA-ABs	Fluorescent treponemal antibody-absorption test
GI	gastrointestinal
G$_M$+	Gram positive
H2S	Hydrogen sulfide gas
HIV	Human immunodeficiency virus
HPV	Human papillomavirus
HSV I	Herpes simplex virus type I
HUS	Hemolytic uremic syndrome
IUGR	Intrauterine growth restriction
LAT	Latency-associated transcript
LFT	Liver function tests
LPS	Lipopolysaccharide
MMR	Measles-mumps-rubella
MRSA	Methicillin resistant *S. aureus*
OLM	Ocular larva migrans
PE	Physical examination
PCP	Pneumocystis pneumonia
PCR	Polymerase chain reaction
PHN	Postherpetic neuralgia
PID	Pelvic inflammatory disease
PMN	Polymorphonuclear

Pt	Patient
RBC	Red blood cell
RES	Reticuloendothelial system
RLL	Right lower lobe
RNA	Ribonucleic acid
rRNA	Ribosomal RNA
RSV	Respiratory syncytial virus
RUQ	Right upper quadrant
SSPE	Subacute sclerosing panecencephalitis
Sx	Symptoms
TCA	Trichloroacetic
TB	Tuberculosis
Tx	Treatment
UG	Urogenitary
VLM	Visceral larva migrans
VSG	Variant surface glycoprotein
WHO	World Health Organization
YO	Year old

Chapter 1
General Principles of Biochemistry

A. An enzyme can be either "stimulated" or "inhibited."
1. "Stimulated" = ↑ in reaction velocity (↑ molecules of product made per unit time).
2. "Inhibited" = ↓ in reaction velocity (↓ molecules of product made per unit time).

B. An enzyme can be "activated" or "deactivated" by another enzyme.

> B (inactive)
> A → ↓
> B* (active)

In the previous example, enzyme B was just activated by enzyme A. In other words, enzyme A catalyzed the reaction that changed B from its inactive to its active form.

C. Receptors can be "up-regulated" or "down-regulated."
1. **Up-regulation** = to increase in number **or** affinity.
2. **Down-regulation** = to decrease in number **or** affinity.
3. High levels of circulating LDL cause a ↓ in the number of LDL receptors (i.e., "high LDL levels cause LDL receptor down regulation").

D. Agents in a pathway can be "upstream" or "downstream."
ex) Pathway: A → B → C → D
1. "Upstream" = reactant molecules that precede the one in reference.
   ex) Molecules A, B, and C are all "upstream reactants" of molecule D.
2. "Downstream" = product molecules produced further down the pathway.
   ex) Molecules B, C, and D are all "downstream products" of molecule A.

E. Rate-limiting step of a reaction pathway—the slowest step.
1. A reaction only proceeds as fast as its slowest step, thus a reaction proceeds only as fast as its rate-limiting step.
2. The rate-limiting step is often the target of regulation because by affecting the speed of this step, the speed of the whole pathway is affected.

F. Oxidation versus reduction
   1. **Oxidation:** Think of it as "tearing apart" or "breaking into smaller pieces" by removing a hydrogen atom from some molecule.
   2. **Reduction:** Think of it as "adding together" or "building up" by adding a hydrogen atom to some molecule.

G. Kinase versus phosphatase enzymes
   1. **Kinase:** Phosphorylates things
   2. **Phosphatase:** DEphosphorylates things

Key idea: Both phosphorylation and dephosphorylation of substrates in cells happens concurrently.

   3. Kinases (e.g., tyrosine kinase or protein kinase A) are phosphorylating substrates <u>at the same time</u> that phosphatases (e.g., protein phosphatase) are dephosphorylating <u>these same substrates</u>. Therefore a dynamic balance exists. The net direction of flux depends on the kinase concentration versus the phosphatase concentration.
      ex) Regulation of glycogen synthesis versus glycogen breakdown is a kinase versus phosphatase phenomenon.
      ex) Regulation of relaxation versus contraction in smooth muscle is a kinase (myosin light chain kinase) versus phosphatase (myosin light chain phosphatase) phenomenon.
   4. **What can ↑ the kinase concentration?** G_s protein linked receptor → ↑ adenylate cyclase → ↑ cAMP → ↑ (kinase)
   5. **What can ↓ the kinase concentration?** G_i protein linked receptor → ↓ adenylate cyclase → ↓ cAMP → ↓ (kinase)

H. Both formation <u>and</u> degradation of cAMP happen at the same time— formation by adenylate cyclase and degradation by phosphodiesterase.

CHAPTER 2
ACID-BASE FUNDAMENTAL CONCEPTS

A. EQUILIBRIUM

1. Equilibrium is the point in time when the **rate of the reverse reaction** (K_2) becomes **equal to the rate of the forward reaction** (K_1).

$$A \underset{K_2}{\overset{K_1}{\rightleftharpoons}} B + C$$

2. Because these rates equalize, once a reaction reaches equilibrium, **there is no change over time in the concentrations of any of the species.**
   ex) After the above reaction reaches equilibrium, the concentrations of A, B, and C will no longer change.

B. ACIDS AND THEIR CONJUGATE BASES

1. An acid (HA) dissociates in water to yield a hydrogen ion (H^+) and a conjugate base (A^-).

$$HA \rightleftharpoons H^+ + A^-$$

2. If the **conjugate base (A^-)** has a **high affinity for H**, it is called a **"strong base."** B/c of its high affinity for H^+, at equilibrium, most of the acid will exist in the **undissociated form (HA^+ and A^-).**
3. Conversely, if the **conjugate base (A^-)** has a **low affinity for H**, it is called a **"weak base."** B/c of its low affinity for H^+, at equilibrium, most of the acid will exist in the **dissociated form (H^+ and A^-).**
4. How "strong" is a certain acid? "Strength" of an acid refers to the extent to which it is dissociated at equilibrium.
5. The extent to which an acid is dissociated at equilibrium (i.e., its "strength") can be represented quantitatively by the Ka (the "acid dissociation constant").

$$Ka = \frac{[H^+][A^-]}{[HA]}$$

6. The previous formula shows us **that the more an acid dissociates into H^+ and A^- at equilibrium, the larger the Ka value will be.** Thus, Ka (and its related term pKa) **express the strength of an acid.**
   pKa = −log [Ka] (thus, a high Ka corresponds to a low pKa)

> **KEY POINTS of the behavior of an acid—conjugate base system**
> - The **less the affinity of the conjugate base for H^+** = the "**weaker**" the **conjugate base** = the **greater** the extent of **dissociation** at equilibrium = the **more H^+** produced in solution = the **higher the Ka** = the **lower the pKa** = the "**stronger**" the acid.
> - The **greater the affinity of the conjugate base for H^+** = the "**stronger**" the **conjugate base** = the **smaller** the extent of **dissociation** = the **less H^+** produced in solution = the **lower the Ka** = the **higher the pKa** = the "**weaker**" the **acid.**

7. By convention, an acid with a **pKa of 2 or less** is considered to be a "**strong acid.**"
   ex) HCL (hydrochloric acid), H_2SO_4 (sulfuric acid)
8. Additionally, an acid with a **pKa of 2 to 6** is considered to be a "**weak acid.**"
   ex) Acetic acid

C. pH (YOU KNOW YOU LOVE IT!)

1. Measures how much H^+ is in a solution (i.e., measures the "acidity" of a solution).
2. Specifically, pH is defined as follows: **pH = −log [H^+]**
3. From this equation, we see that…
   a. The **higher the H^+ concentration** in the solution = the **lower the pH** = the **more "acidic."**
   b. The **lower the H^+ concentration** in the solution = the **higher the pH** = the **more "basic."**
4. By definition…
   a. A solution with an **[H^+] of 10 to 7** has a **pH of 7** and is called a **neutral solution.**
   b. A solution with an **[H^+] greater than 10 to 7** has a **pH of less than 7** and is called an **acidic solution.**
   c. A solution with an **[H^+] less than 10 to 7** has a **pH greater than 7** and is called an **alkaline solution.**

D. BUFFERS (NOT JUST FOR POLISHING YOUR CAR!)

1. A buffer is a solution that has a **mixture of a weak acid and its conjugate base.** The goal of a buffer is to maintain the pH of a solution at a constant level.
   a. Conceptual example: $\quad HA \rightleftharpoons H^+ + A^-$

   b. Physiologic example: $\quad H_2CO_3 \rightleftharpoons H^+ + HCO_3^-$
      (carbonic acid)   (biocarbonate)

2. **How does a buffer maintain the pH of a solution?**
    a. B/c it has **both** acid **and** base present, the buffer can counteract any attempt to change the pH.
        ex) If base is added to the solution, the buffer system responds by using some of its weak acid (HA) to donate H^+ to neutralize the base
        ex) If acid is added to the solution, the buffer system responds by using some of its base to absorb the H^+ released by the acid.

KEY IDEA: Therefore we see that the buffer system acts to mitigate the potentially destabilizing effects of adding either base or acid to a system.

    b. At a given pH, the best buffer will have <u>equal concentrations</u> of both weak acid and conjugate base. (This allows it to "defend" equally well against <u>either</u> the challenge of added acid <u>or</u> added base.)
3. **How do you pick the best buffer for a given pH?** (That is, if I want to maintain the pH of a solution at 7, which weak acid would I pick to act as a buffer?)
    a. Short answer: Pick a weak acid with a **pKa equal to the desired pH.**
    b. Long answer: Let us examine a titration curve for phosphoric acid (a weak acid) to understand why this is so (Figure 2.1).

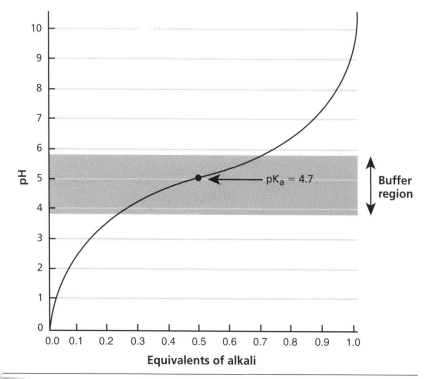

FIGURE 2.1 Titration curve.

4. **Where is this weak acid best functioning as a buffer?**
   a. The best buffering region = the place where pH changes the least in response to a given amount of base added = the flattest part of the curve
      - Note: **The flattest part of the curve occurs near the pKa.**
   b. Therefore if we wanted to keep a solution at a pH of 7, phosphoric acid (with a pKa of 6.7) would be great because its best buffering capacity occurs at pH values near 7.

E. **THE PHYSIOLOGIC BUFFER SYSTEM: YOU COULDN'T LIVE WITHOUT CO_2 ←→ H_2CO_3 ←→ HCO_3-**
   1. **The carbon dioxide (CO_2) – carbonic anhydrase (H_2CO_3) – bicarbonate (HCO_3-) buffer system is the body's most important mechanism for maintaining pH near 7.**
   2. All three substances exist in equilibrium and can be interconverted as follows:

$$CO_2 \underset{\longrightarrow}{\longleftarrow} H_2CO_3 \underset{\longrightarrow}{\longleftarrow} H^+ + HCO_3-$$
$$\uparrow \text{carbonic anhydrase}$$

   3. **How does this system work?**
      a. **Carbonic acid (H_2CO_3) functions as the weak acid.** If the pH becomes too alkaline, carbonic acid counteracts this change by dissociating to produce H^+.

$$H_2CO_3 \longrightarrow H^+ + HCO_3-$$

      b. **Bicarbonate (HCO_3-) functions as the conjugate base.** If the pH becomes too acidic, bicarbonate counteracts this change by reacting with H^+ to remove it from circulation.

$$H_2CO_3 \longleftarrow H^+ + HCO_3-$$

   4. **What happens if the body's supply of carbonic acid gets too low?** It **uses CO_2 to replenish the supply,** through the carbonic anhydrase reaction:

$$CO_2 \underset{\longrightarrow}{\longleftarrow} H_2CO_3$$
$$\uparrow \text{carbonic anhydrase}$$

   5. **What happens if the body's supply of bicarbonate gets too low?** → The **kidney reabsorbs more** of it.

HPI: 20 yo man presents with confusion, abdominal pain, rapid breathing, vomiting, and diarrhea of acute onset. The patient had been drinking wine the night before, when he passed out. On PE, blood pressure is 88/48 and tachypnea is noted ~28 breaths per minute. The patient's mucosal membranes are dry, and a lack of lacrimation is noted. The skin is dry, cold, and slightly blue. A particular "fruity" odor emanates from the patient. Arterial blood gas (ABG) shows reduced bicarbonate and an anion gap.

Diabetic Keto**acidosis**. Ketoacidosis is most common in people with type 1 diabetes. Ketone, acetoacetate, and Beta-hydroxy-butyrate production occur during inability to metabolize alcohol, and following times of stress, infection, and exertion. Hyperglycemia is present, as is glycosuria and ketonuria.

The pH is too low (a situation called **"acidosis"**).
How does the body respond?
1. ↑ **buffering activity of bicarbonate** to the reduce H^+ levels
2. ↑ **bicarbonate reabsorption** in the kidney (to replenish the supply)
3. ↑ **H^+ secretion** from the kidney into the urine
4. ↑ **CO_2 exhalation**; reducing levels of CO_2 "draws" both reactions to the left, consuming H^+ in the process.

HPI: 4 yo presents with persistent vomiting × 2 weeks. The family has been reluctant to bring the child in for medical attention for various reasons, and apparently the child is severely dehydrated. On PE, the skin is dry, skin turgor is poor, the eyes are sunken, and lacrimation and sweating are absent. Severe malaise and muscle weakness is evident. Complete blood count (CBC) shows an increased hematocrit. A Chem 7 shows hypokalemia. The ABG reveals...?

Metabolic Alkalosis. **The pH is too high** (a situation call **"alkalosis"**).
How does the body respond?
1. ↑ **buffering activity of the carbonic acid** to produce H^+ ($H_2CO_3 \rightarrow H^+ + HCO_3^-$)
2. ↓ **bicarbonate reabsorption** in the kidney (b/c lowering bicarbonate levels "encourages" the carbonic acid buffering reaction we just mentioned)
3. ↓ **H^+ secretion** from the kidney into the urine
4. ↓ **CO_2 exhalation** (conserving CO_2 helps replenish the supply of carbonic acid)

CHAPTER 3

DNA AND RNA—STRUCTURE AND IMPORTANT PROCESSES (REPLICATION, TRANSCRIPTION, AND TRANSLATION)

A. DNA: WHAT IS IT? WHO CARES?

Basic definition: Genetic material that is passed on from parent cell to daughter cell and that carries the information necessary to build and direct the functioning of all parts of the cell (Figure 3.1).

B. NITROGENOUS BASES (N BASES): WHAT YOU NEED TO KNOW

1. Three of these N bases have **one ring**—these are the three **"pyrimidines"** (Figure 3.2).
    a. Cytosine (C)
    b. Uracil (U)
    c. Thymine (T)
    d. **Cheesy little mnemonic → CUT the PY** (pie)
2. Two of the N bases have **two rings**—these are the two **"purines"** (Figure 3.3).
    a. Adenine (A)
    b. Guanine (G)

> **KEY IDEA: How is information encoded in DNA?** Based on **the sequence of the N bases in a strand.** The sequence of N bases ultimately determines the sequence of amino acids, which, in turn, determines the structure and properties of a protein.

C. NUCLEOTIDES: BASIC STRUCTURAL UNIT OF DNA AND RNA: A MOLECULE WITH THREE IMPORTANT PARTS

1. **The sugar has a free 3' OH group:** This group can bond with the 5' triphosphate group of an incoming nucleotide to link them together via **phosphodiester bonds** (Figure 3.4).
2. **The nitrogenous base:** This base can form hydrogen bonds with other N bases.
3. **The 5' triphosphate group:** This group **forms phosphodiester bonds** with the 3' OH group of another nucleotide.

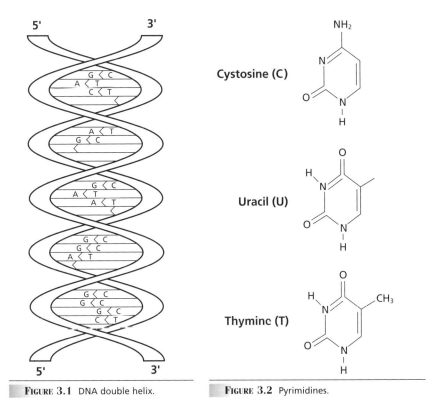

FIGURE 3.1 DNA double helix.

FIGURE 3.2 Pyrimidines.

D. STRUCTURE OF DNA

1. DNA is a double stranded helix—two strands intertwined with each other.
2. **How are the strands held together?**
   By hydrogen bonds between nitrogenous bases.
   a. A forms two H bonds with T or U
      A = T / U
   b. G forms three H bonds with C
      G ≡ C

KEY IDEA: A G ≡ C bond (three H bonds) is stronger than an A = T bond (two H bonds).

3. Each DNA strand consists of a bunch of nucleotides linked together (Figure 3.5).
4. **How are the nucleotides linked together to form a strand?**
   By **phosphodiester bonds** that form between the 5' phosphate group of one nucleotide and the 3' OH group of another one. These bonds are covalent and therefore strong (Figure 3.6).

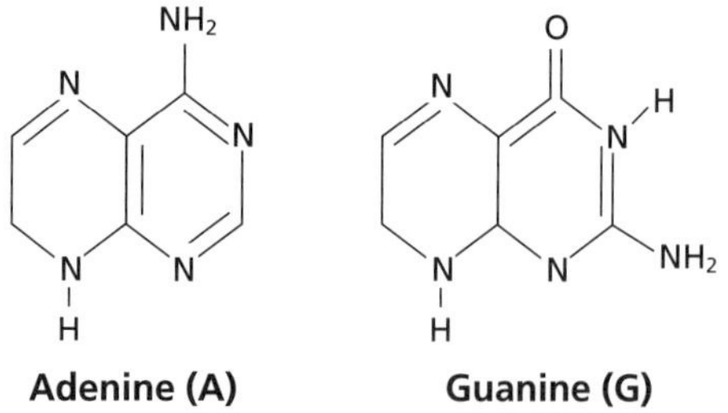

NH2
N
N
N
N
H
Adenine (A)
O
N
H
N
N
N
NH2
H
Guanine (G)

FIGURE 3.3 Purines.

O
O
P
O
C
H
O⁻
H
Phosphate
Group
Base
Sugar
OH
OH

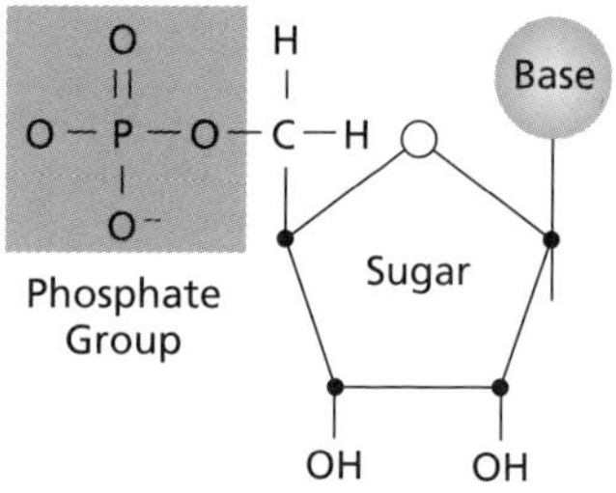

FIGURE 3.4 Nucleotide.

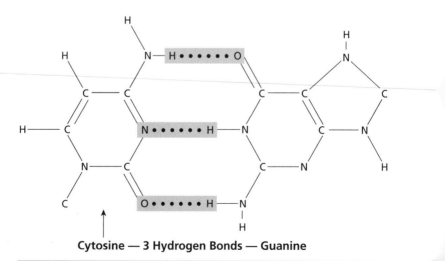

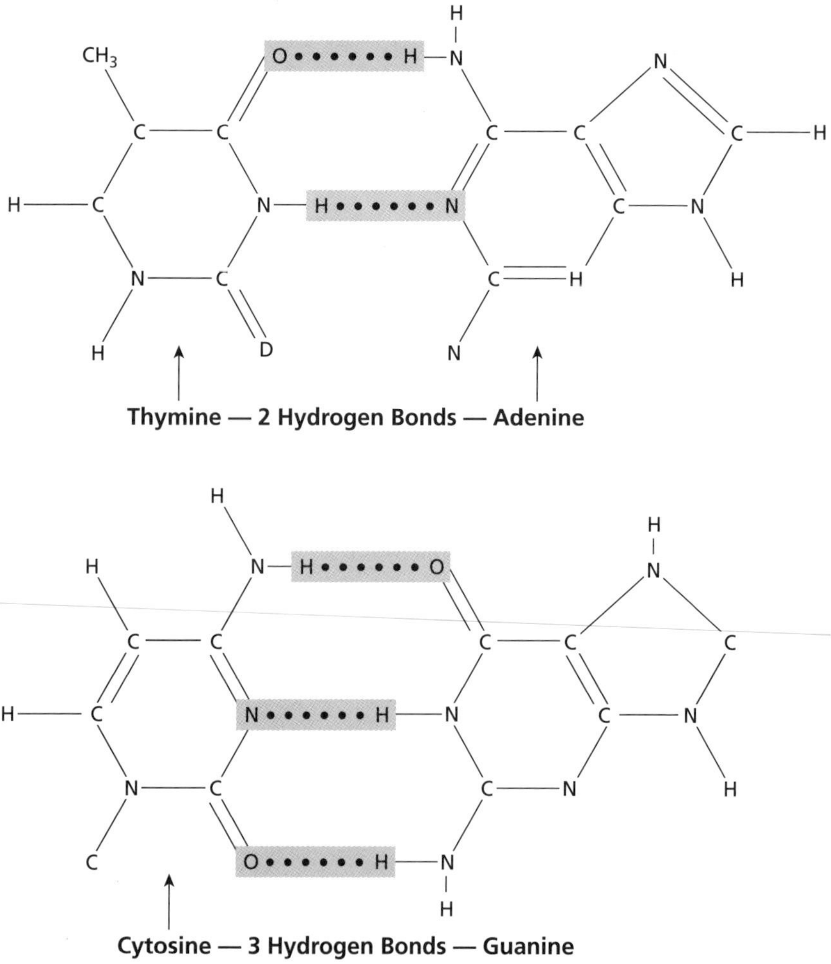

Thymine — 2 Hydrogen Bonds — Adenine
Cytosine — 3 Hydrogen Bonds — Guanine

FIGURE 3.5 Base pairing.

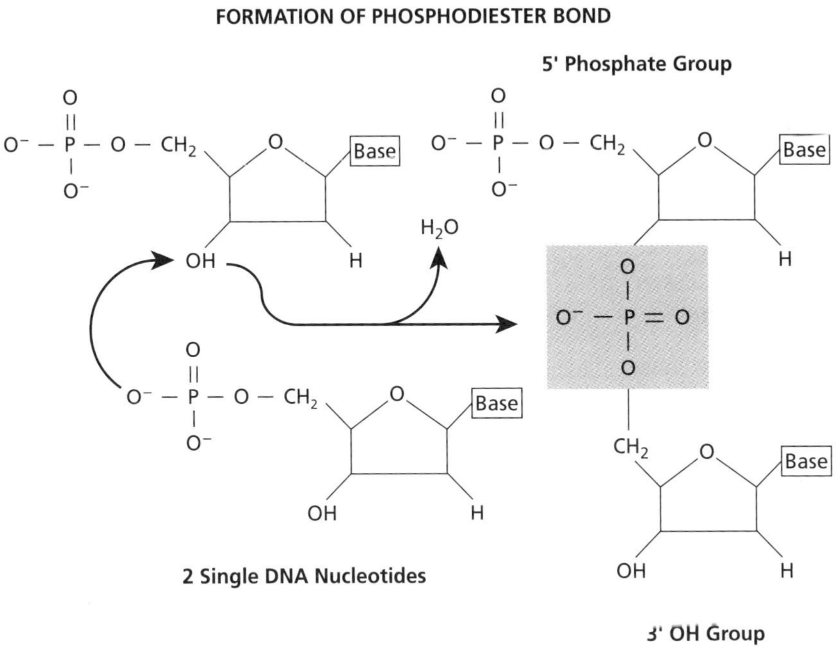

DNA AND RNA—STRUCTURE AND IMPORTANT PROCESSES

FORMATION OF PHOSPHODIESTER BOND

FIGURE 3.6 Structure of phosphodiester bond.

> **KEY IDEA:** If a nucleotide does not have a 3' OH group, then it can**NOT** form phosphodiester bonds with other nucleotides, and a DNA strand canNOT be formed.

5. Clinical correlation → Many drugs that block DNA and RNA synthesis work because they are **nucleotide analogues that lack a 3' OH group.** Thus, when incorporated into a newly forming DNA or RNA strand, they are **unable to form phosphodiester bonds.** The result that the strand cannot be elongated any further (i.e., "**chain termination**").
   ex) **Nucleoside reverse transcriptase inhibitors (NRTIs) used in HIV therapy.** Because they lack a 3' OH group, these nucleotide analogs inhibit the function of reverse transcriptase.
6. Can you name five examples of NRTIs?
   a. Zidovudine (AZT)
   b. Didanosine (ddI)
   c. Zalcitabine (ddC)
   d. Stavudine (d4T)
   e. Lamivudine (3TC)
   ex) **Antivirals used against the herpes family** (e.g., HSV, CMV, EBV, VZV). Because they lack a 3' OH group, these nucleotide analogs inhibit DNA replication and RNA transcription. Name the two most important drugs of this type.

- Acyclovir (requires viral thymidine kinase for activation)
- Ganciclovir (does NOT require viral thymidine kinase for activation)

ex) The anti-cancer drug cytosine arabinoside

7. More **properties of the genetic code**
   a. It is **contiguous** (i.e., codons do not overlap, nor are they separated (except for some viruses).
   b. It is **unambiguous** (i.e., each codon specifies **only one** amino acid).
   c. It is **degenerate** (i.e., some amino acids are specified by more than one codon).
   d. It is **universal** (i.e., a certain codon codes for the same amino acid, regardless of whether we are dealing with a plant, animal, or bacterial DNA).
      Exceptions: Mitochondria, mycoplasma, Archaeobacteria, and some yeasts.

E. STRUCTURE OF RNA

RNA is also made from nucleotides and is identical to DNA (Figure 3.7), EXCEPT in the ways shown in Table 3.1.

F. STRUCTURE OF A GENE

1. **Gene (simplified definition):** a section of DNA that codes for a certain protein (Figure 3.8).
2. **What about sequences of DNA that do not code for proteins?**
   a. Some of them are <u>promoter regions</u> (e.g., the TATAAT sequence in both eukaryotes and prokaryotes).
   b. Some are <u>binding sites for regulatory proteins</u>.
   c. Some are <u>signal sites for gene rearrangements</u>.

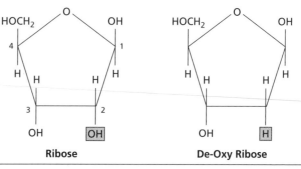

FIGURE 3.7 RNA versus DNA.

TABLE 3.1 Differences between RNA and DNA

RNA	DNA
Single stranded	Double stranded
Sugar backbone uses ribose	Sugar backbone uses deoxyribose
Uracil (instead of Thymidine)	Thymidine

G. Overview: Key Ideas of Replication, Transcription, and Translation
(Table 3.2)

H. DNA Replication (DNA→DNA) (in the nucleus)
1. **Steps in the process of DNA replication** (Figure 3.9)
   a. DNA helicase unzips and unwinds the strands
   b. RNA primer binds
   c. DNA polymerase III begins chain formation by adding a nucleotide to the 3' OH group of the RNA primer
   d. DNA polymerase I deletes RNA primers and fills those spaces with nucleotides
   e. DNA ligase joins all strands together (Figure 3.10)

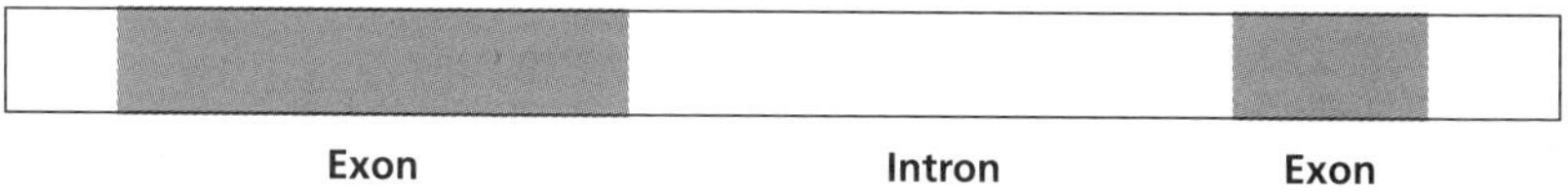

| Exon | Intron | Exon |

Figure 3.8 Structure of a gene.

Table 3.2 Key Ideas of Replication, Transcription, and Translation

Context	Process	Location	Key Players
Cell is making a copy of itself; therefore it needs to copy its DNA	Replication DNA → DNA	Nucleus	• RNA primer • DNA polymerase III • DNA ligase
Cell needs to make some proteins (2-step process)	Step 1. Transcription DNA → mRNA	Nucleus	• Promoter region • General transcriptional factors • RNA polymerase
	Step 2. Translation mRNA → protein	Cytoplasm	• mRNA strand • Ribosomes • tRNAs • Amino acids

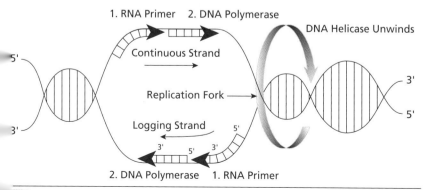

Figure 3.9 DNA replication—an overview.

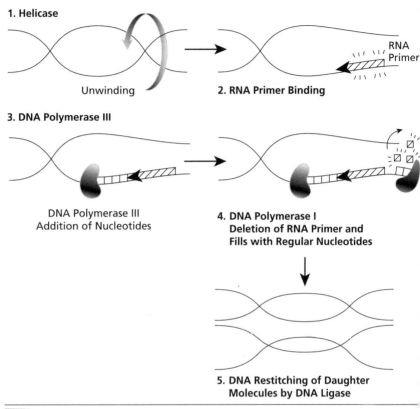

FIGURE 3.10 Steps of DNA replication.

2. **DNA replication—name that enzyme!** (Table 3.3)
3. **Clinical correlation** → Certain anticancer drugs inhibit DNA replication by **cross-linking DNA**. If the strands are cross-linked, then DNA helicase cannot unzip and unwind them, and replication is thwarted (see Table 3.3).
   ex) Cisplatin
   ex) Nitrosureas
   ex) Cyclophosphamide
4. **Clinical correlations:**
   → The anticancer drug **etopo**side inhibits DNA replication by **inhibiting topo**isomerase 2 (the strain of unwinding the helix is not relieved; DNA strand breakage results).

Table 3.3 DNA Replication—Name that Enzyme!

Function	Enzyme
Unzips and unwinds strands?	DNA helicase (using energy from ATP hydrolysis)
Reversibly cuts strand of helix to relieve strain imposed by unwinding?	Topoisomerases (eukaryotes) DNA gyrase (bacteria)
Holds the strands apart?	Single stranded binding protein
Makes the RNA primer?	Primase
Begins replication by adding nucleotides to 3' OH group of RNA primer?	DNA polymerase III (using energy provided by cleaving pyrophosphate from each nucleotide)
"Proofreads" each nucleotide? (If an error is found, it is excised and replaced.)	3'-5' exonuclease activity of DNA polymerase III
Deletes many RNA primers used on lagging strand?	3'-5' exonuclease activity of DNA polymerase I
Fills in gaps left by deleted RNA primers on lagging strand?	DNA polymerase I
Ligates different pieces of lagging strand together?	DNA ligase

The antibacterial drug class **fluoroquinolones** inhibit bacterial DNA replication by **inhibiting DNA gyrase** (the strain of unwinding the helix is not relieved; DNA strand breakage results) (see Table 3.3). Name three important fluoroquinolones....
  a. Cipro**floxacin**
  b. O**floxacin**
  c. Nor**floxacin**

5. **Clinical correlation**
   → This antiviral drug blocks DNA replication by inhibiting viral DNA polymerase III.
   What is this antiviral drug? → Foscarnet.

6. **Clinical correlation**
   → Three anticancer drugs use **DNA intercalation** as one of their strategies to inhibit DNA replication. By inserting themselves into the DNA strand, they inhibit the work of DNA polymerase III (see Table 3.3). What are these three drugs?
   a. Doxorubicin
   b. Dactinomycin
   c. Bleomycin

KEY IDEA: DNA replication:
- **When does DNA replication begin?**
  → When an RNA primer binds.
- **Why is an RNA primer necessary?**
  → B/c DNA polymerase III can add only nucleotides to the 3' OH group of an existing structure. (The RNA primer—with a free 3' OH group—provides such a structure.)
- **Eukaryotic cells** have **DNA replication** beginning **from many points simultaneously** (i.e., "multiple origins of replication"). HOWEVER, **bacteria, viruses and plasmids** have **only one origin of replication.**
- **How does DNA replication proceed?**
  → DNA polymerase III binds and starts adding nucleotides to the RNA primer.

HUGE KEY IDEA: "DNA replication proceeds from 5' to 3' on the newly synthesized strand." (Say it over to yourself a few times and let it sink in. It is key!)
- **Why can it only proceed from 5' to 3' on the newly synthesized strand?**
  → B/c DNA polymerase can only add new nucleotides at one place—the 3' OH group of an existing nucleotide (this is the only site for phosphodiester bond formation).
- **Leading strand versus lagging strand**
  - **Leading** strand: Synthesized in **one continuous piece**
  - **Lagging** strand: Synthesized as a series of fragments (**"Okazaki fragments"**), which are then **ligated (linked) together by DNA ligase**
- **When does DNA replication end?**
  → When the different replication forks meet up with each other.

I. **TRANSCRIPTION (DNA → RNA) (IN THE NUCLEUS)**
  1. **Steps in the process of transcription** (Figure 3.11)
     a. **Binding:** General transcriptional factors and RNA polymerase bind to promoter site.
     b. **Initiation:** RNA polymerase begins chain formation (no primer needed).
     c. **Elongation:** RNA polymerase continues chain formation (5' → 3' on the newly synthesized strand).
     d. **Termination:** RNA polymerase stops chain elongation when some termination sequence of nucleotides is reached.

KEY IDEA: The mRNA ends up looking just like the SENSE strand of DNA, except that Uracil is substituted for Thymidine.

  2. **Clinical correlation: Drugs that work by preventing transcription of RNA**

KEY IDEA: Most of the drugs that block DNA replication also block RNA transcription! In other words, because of their mechanisms, these drugs block both DNA replication and RNA transcription.

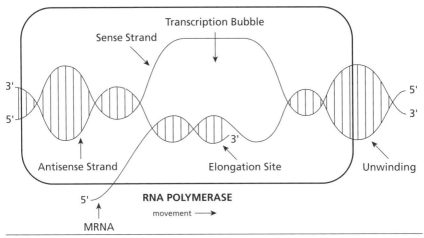

FIGURE 3.11 Transcription.

    a. MOA → **Cross-link DNA.** (If the strands are cross-linked, then they cannot be unwound and unzipped. Thus both replication and transcription are thwarted.)
       ex) Anticancer drugs cisplatin, nitrosureas, and cyclophosphamide
    b. MOA → **Nucleotide analogs that lack a 3' OH group.** They cause chain termination in the growing DNA or RNA strand.
       ex) The anticancer drug **cytosine arabinoside**
       ex) The antivirals used against the herpes family—**acyclovir and gangcyclovir**
    c. MOA → **DNA strand breakage due to unrelieved strain.** (A broken strand obviously impairs both replication and transcription)
       ex) The anticancer drug etoposide (**inhibits topo**isomerase 2)
       ex) The antibacterial drug class of **fluoroquinolones (inhibit DNA gyrase)**
    d. MOA → **DNA intercalation.** By inserting themselves into the DNA strand, they inhibit the work of both DNA polymerase and DNA dependent RNA polymerase
       ex) The anticancer drugs doxorubicin, dactinomycin, and bleomycin
    e. The only drug that blocks RNA transcription, but does NOT also interfere with DNA replication, is the antibacterial drug **rifampin**. It inhibits bacterial RNA transcription by **binding to DNA dependent RNA polymerase and inhibiting it**.
3. **What happens to an RNA strand after it is synthesized?**
   → It undergoes **posttranscriptional modifications** (aka **RNA processing**) in the **nucleus**.
    a. **Poly Adenylate tail** added to 3' end...**poly Guanine cap** added to 5' end
       • **Why the poly G cap?** → To protect RNA from endonuclease digestion; you have a buffer of "extra" nucleotides that can absorb any potential damage.

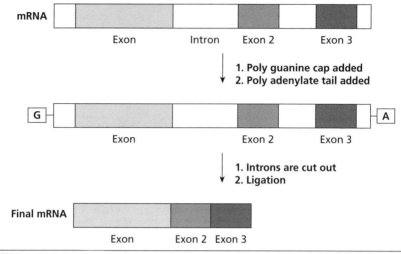

FIGURE 3.12 RNA processing.

    b. **Introns are cut out**
       • Introns = noncoding region
       • Exons = coding regions
    c. **Exons are ligated together**
       • **Mnemonic** → **IN**trons stay **IN** the nucleus; **EX**ons **EX**it and are **Ex**pressed (Figure 3.12)

KEY IDEA: Only processed RNA is transported out of the nucleus.

J. THREE DIFFERENT STRUCTURES MADE FROM RNA

KEY IDEA: mRNA isn't the only thing that RNA is used for. **In fact, there are three different structures that are made from RNA strands.**
1. **mRNA (messenger RNA):** A "copy" of a certain section of DNA.
   • mRNA is the largest of the three structures made from RNA.
2. **tRNA (transfer RNA):** The **little structures** that float in the cytoplasm and are **used in protein production**. One side has an anticodon; the other side carries a certain amino acid.
   • tRNA is the **smallest** of the three structures made from RNA.
3. **rRNA (ribosomal RNA):** This RNA gets used as the **structural component of ribosomes**.
   • rRNA is the **most abundant** of the three structures made from RNA (Figure 3.13).

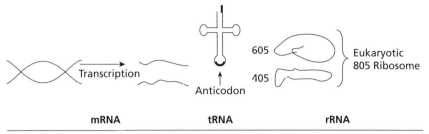

FIGURE 3.13 Types of RNA.

KEY IDEA: The process of **"transcription"** does **NOT** just refer to making mRNA; it refers to making <u>any</u> RNA strand.

K. REGULATION OF TRANSCRIPTION

How does the body ↑ transcription of a certain gene? (2 ways)
1. **↑ Speed of assembly of general transcriptional factors** at the promoter site
   - <u>Faster assembly</u> of general transcription factors = <u>less time required</u> before each RNA polymerase can bind = <u>more transcription</u> of the gene per unit time.
2. Has an **"activator protein"** bind at a site from which it **physically facilitates the binding of RNA polymerase** to the promoter site, thus facilitating the initiation of transcription?
   - The site at which it binds is called an **"enhancer site."**

KEY IDEA: **This enhancer site can be thousands of base pairs away** from the promoter region and yet still exert its effect. **How?** DNA looping. (DNA loops around to bring the two sites into proximity.)

- **Clinical correlation → Steroid hormones act like activator proteins.** They facilitate the binding of RNA polymerase to the promoter site; thus ↑ transcription of a certain gene.

How does the body ↓ transcription of a certain gene? (3 ways)
1. **↓ Speed of assembly of general transcription factors** at the promoter site
   - <u>Slower assembly</u> of general transcription factors = <u>more time required</u> before each RNA polymerase can bind = <u>less transcription</u> of that gene per unit time.
2. Have a **"repressor protein"** bind at a site from which **it physically prevents RNA polymerase from binding** to the promoter site and beginning transcription.

KEY IDEA: This site can be thousands of base pairs away from the promoter region and yet still exert its effect. **How?** DNA looping. (DNA loops around to bring the two sites into proximity.)

   3. Use a repressor protein to inhibit the function of the activator proteins
L. TRANSLATION (mRNA → PROTEIN) (IN CYTOPLASM) (Figures 3.14 and 3.15)
   1. Steps in the process of translation
      a. **Initiation.** Ribosomal subunits, initiation factors, GTP, and a tRNA—carrying methionine—bind at the AUG codon on the 5' end of the mRNA strand

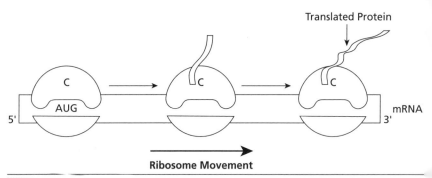

FIGURE 3.14 Translation—overview.

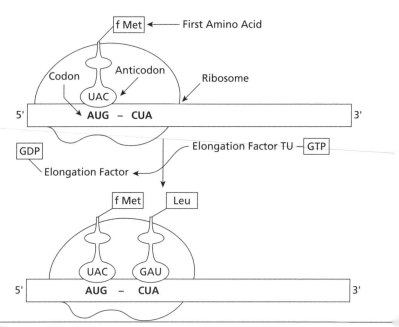

FIGURE 3.15 Translation—steps.

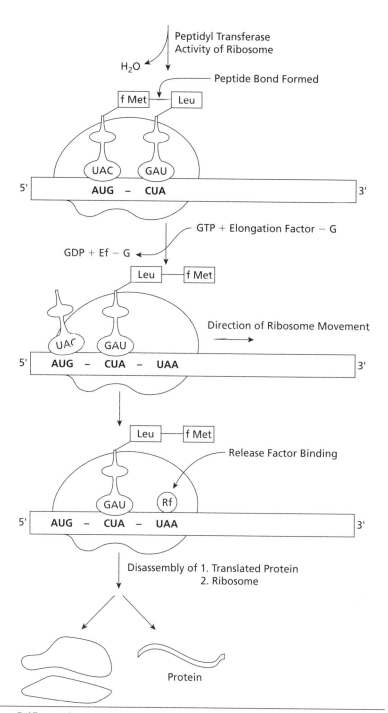

FIGURE 3.15—CONT'D Translation—steps.

b. **Elongation.** Three-step cycle:
   (1) An incoming tRNA binds at the A site
   (2) The peptide chain is transferred from the P site to the A site and a new peptide bond is formed (enzyme = peptidyl transferase)
   (3) The ribosome shifts to the right—kicking out the tRNA that was in the P site, and creating a new, vacant A site.
c. **Termination.** Ribosome encounters a stop codon (UGA, UAG, UAA). This causes release of the newly synthesized peptide chain and dissolution of the ribosomal complex.
2. **As translation proceeds—key points you need to know**
   a. A certain **tRNA enters the A site and binds based on complementary base pairing (G-C, A-U) between its codon and the mRNA codon** exposed in the A site.
      - Note: The codon on a tRNA is called an "anti codon."
   b. The amino acids on the growing chain **are linked N terminus to C terminus**
   c. **Many ribosomes proceed down the mRNA strand at the same time**—each in a different stage of completion of the protein.
      - This speeds up the process of protein production.
3. **Clinical correlation → Certain antibacterial drugs inhibit protein synthesis by binding to one of the ribosomal subunits and inhibiting translation.**
   ex) 30s inhibitors
   a. Aminoglycosides (bacteriocidal)...irreversibly bind; prevent formation of initiation complex.
   b. Tetracyclines (bacteriostatic)...bind and prevent tRNA binding at the A site.
   ex) 50s inhibitors (all are bacteriostatic)
   - Chloramphenicol. Binds and prevents the action of peptidyl transferase
   - Erythromycin. Binds and interferes w/translocation.
   - Lincomycin. Binds and interferes w/translocation.
   - Clindamycin. Binds and interferes w/translocation.

M. **SUMMARY AND COMPARISON OF REPLICATION, TRANSCRIPTION, AND TRANSLATION**
(Table 3.4)

N. **TYPES OF DNA MUTATIONS AND HOW THEY AFFECT PROTEIN STRUCTURE**
   There are two broad categories of mutations:
   1. **Point mutations:** One nucleotide gets substituted for another
      a. A point mutation can create a <u>different codon</u>, <u>but</u> the <u>same amino acid</u> still gets incorporated into the protein (since some amino acids are coded for by more than one codon). This type of mutation is called what?
         → A **"silent mutation"**
      b. A point mutation can create a <u>different codon</u>, and the result is that a <u>different amino acid</u> gets incorporated into the protein. This type of mutation is called what?
         → A **"missense mutation"**

TABLE 3.4 Comparison of Replication, Transcription, and Translation

	DNA → DNA (replication)	DNA → RNA (transcription)	mRNA → protein (translation)
Location	Nucleus	Nucleus	Cytoplasm
Initiation?	RNA primer binds	General transcriptional factors and RNA polymerase bind to promoter sequence (TATAAT)	Two ribosomal subunits bind at universal start codon (AUG) at 5' end of mRNA strand
Elongation?	• DNA polymerase III links nucleotides together • Proceeds 5' to 3' on newly synthesized DNA strand	• RNA polymerase links nucleotides together • Proceeds 5' to 3' on newly synthesized RNA strand	• Ribosomal subunits slide along mRNA strand; tRNAs read codons and attach amino acids • Proceeds 5' to 3' along mRNA strand • Amino acids link N terminus to C terminus
Termination?	Different replication forks meet each other	Some termination sequence of nucleotides is reached	Stop codon is reached (eukaryotes = UAG, UAA, or UGA)
How does the cell accelerate the process?	Multiple replication forks all work on same chromosome at the same time (eukaryotes only) • Bacteria, viruses, and plasmids have only one replication fork	Multiple RNA polymerases all travel down DNA and transcribe at the same time	Multiple ribosomal complexes all travel down mRNA strand at the same time

  c. A point mutation can create a <u>premature stop codon</u> in the middle of the mRNA strand. The result is truncated translation, and a half-formed protein. This type of mutation is called what?
  → A **"nonsense mutation"**
2. **Insertions or deletions of a nucleotide**
  a. Insertion or deletion of a nucleotide is the mutation that <u>is most damaging to protein production</u> b/c it can cause misreading of all downstream codons. The result is a slew of misplaced amino acids. This type of mutation is called what? →
  A **"frameshift mutation"**
  b. **Still unclear about frameshift mutation?** Remember that an mRNA strand has a string of N bases, which are read in groups of three (called "codons") by tRNA. If one N base in missing, or an extra one is added, what does the tRNA do? It starts with the first 3 N bases it "sees", calls these a codon, and then subdivides the rest of the strand into new groupings of three. Thus all downstream N bases are affected b/c they are "subdivided" into new groups of three (Figure 3.16).

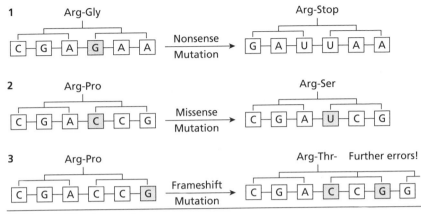

FIGURE 3.16 Point mutations.

KEY IDEA: Severity of damage: Frameshift > nonsense > missense > silent

O. CLINICAL CORRELATIONS

Mutations are bad. Here are five disparate examples from different fields to demonstrate the wide-ranging effect of changing a few bases here and there. Mutations will result in either a **decrease** in expression of a necessary gene, or **excessive** expression, often leading to cancer.

HPI: 8 yo girl of Irish descent presents with cough, fever, and green expectorate (the slime from coughing). She has had a history of foul-smelling stools and recurrent bladder infections. On PE, the patient is tachypnic (breathing fast), tachycardic (heart beating fast), and hypertensive (high blood pressure). Nasal polyps are noted along with a barrel-shaped chest. Stool test reveals significant lipids. A sweat test is positive for excessive Na (sodium) production. What is the diagnosis?
→ Cystic fibrosis. The classic mutation!
- Autosomal recessive mutation of CFTR (cystic fibrosis transmembrane conductance regulator) gene: A mutation in the long arm of chromosome 7, band q31 → chloride transport out of lumen of sweat gland slowed → excess salt secretion.

HPI: 7 yo boy presents with dislocated right shoulder. He was wrestling with his brother when he heard something "pop." This is the fifth time this year that something such as this has happened. The parents adamantly deny any child abuse. On PE, bruises and scars are noted on the knees and elbows.

Continue

Bilateral sclera have a bluish tint. CBC and Chem 7 (blood work) are normal. A clotting profile of PT/PTT (coagulation times to assess the blood's ability to clot) is normal as well. The astute medical student makes the diagnosis of...?
→ Ehlers-Danlos syndrome.
- Faulty collagen synthesis secondary to various mutations. Different types have different transmissions.

HPI: 24 yo man complains of infertility. His wife has been checked and given a clean bill of health. His sperm count is normal, but abnormalities were noted on the motility of the sperm. He comes to you for a second opinion. His medical history in noteworthy for frequent upper respiratory infections. You put two and two together to give a diagnosis of...?
→ Kartagener's syndrome.
- Autosomal recessive mutation resulting in lack of dynein → immotile sperm and immotile cilia (URIs).

HPI: 24 yo "Dracula" presents with a lesion on his hand that has changed in size and color recently. He has avoided sunlight at all costs because of photosensitivity. On PE, numerous telangiectasias are present along with freckles on the face. Hyperkeratosis is present on the hands and elbows. Biopsy of the lesion gives the diagnosis of basal cell carcinoma. The underlying predilection to the cancer is...?
→ Xeroderma pigmentosum.
- Autosomal recessive disorder → impaired endonuclease function of DNA repair mechanisms → skin cancer. All cancers begin with mutations. When mutations exceed the cell's capability to repair them, expression of the defective genes occurs.

HPI: 38 yo woman presents to the ER after being found in a coma. She had complained of dizziness when she went without eating for prolonged durations. On PE, tachycardia is noted along with cold, clammy and sweaty hands. MRI (magnetic resonance imaging) shows a mass in the tail of the pancreas. An infusion of glucose aids the patient out of her coma. Elevated immunoreactive C-peptide is noted. What is the diagnosis?
→ Insulinoma.
- B-cell pancreatic islet cell tumor. Can be associated with MENs (Multiple Endocrine Neoplasia). Errors in translation result from **cleavage** of newly made protein chains, **defective targeting** of proteins, or from mutations in the original genes. In the case of an insulinoma, a **mutation** results in the increased precursor of preproinsulin and eventually excess insulin. In the case of a familial hyperproinsulinemia, elevated levels of proinsulin result from defects in protein **cleavage** of the newly translated peptide-preproinsulin. (Protein cleavage is discussed in the next chapter.)

CHAPTER 4
TECHNIQUES OF BIOTECHNOLOGY AND MOLECULAR BIOLOGY

A. PROCEDURES INVOLVING DNA AND/OR RNA
   1. **How do you create fragments of DNA (for use in analysis)?**
      → With restriction endonucleases (also known as restriction enzymes), which cleave DNA wherever certain palindromic sequences are present.
   2. **What is RFLP (restriction fragment length polymorphism)?**
      a. When you cut DNA with a certain restriction endonuclease, the fragments you produce are called **"restriction fragments."** **"Polymorphism"** refers to the fact that restriction fragments can have different lengths.
      b. **The key idea is that small differences in DNA sequence can produce restriction fragments of different lengths. How?**
         → By altering the restriction enzyme cutting pattern
      c. Thus, after cleaving the same section of DNA from two different people with the same restriction enzyme, scientists can compare the lengths of the restriction fragments and tell how genetically similar or dissimilar those people are.
      d. Restriction fragments of <u>identical lengths</u> indicate <u>genetic similarity</u>.
      e. Restriction fragments of <u>different lengths</u> indicate differences in DNA sequence thus <u>genetic dissimilarity</u>.
      f. RFLP analysis has been used to help construct the map of the human genome. It is also used by anthropologists to assess the degree of genetic similarity between different groups of people.
   3. **What is cDNA and why is it useful?**
      a. Complimentary DNA is "reverse-engineered" from an mRNA strand using reverse transcriptase. Once you do this you get a strand of DNA that includes only the coding regions (i.e., no introns and no regulatory sequences).
      b. Uses of cDNA:
         (1) Amplified by cloning or PCR for use in experiments
         (2) As a probe to locate the gene that codes for the sample mRNA strand from which the cDNA was made

B. You need to understand four procedures involving proteins

1. **Sequencing using the dideoxynucleotide (ddNTP) chain termination method**
   a. **Goal:** To determine the sequence of bases of a certain DNA sample
   b. **Logic:** Have the unknown DNA fragment undergo replication in the presence of four different radiolabeled dideoxynucleotides (ddATP, ddGTP, ddCTP, ddTTP). Each time a ddNTP is incorporated into the chain, the chain is terminated (because ddNTPs have no 3' OH group). The resulting fragments of different lengths can be compared and the base pair sequence of the original sample deduced.
   c. **Steps:**
      (1) Denature the unknown DNA fragment into single strands, and divide these strands among four different samples (one sample for each of the four types of nucleotides).
      (2) To each sample add: Primer, DNA polymerase, one type of nucleotide, and that same type of dideoxynucleotide. Allow replication to proceed.
      (3) Once replication is done, subject the fragments to autoradiography and then visualize the band patterns. By reading these patterns correctly, the sequence of the original fragment can be deduced.

2. **PCR (polymerase chain reaction)**
   a. **Goal:** Produce many copies of a desired DNA fragment
   b. **Logic:** Take a DNA fragment and repeatedly subject it to replication.
   c. **Steps:**
      (1) **Denaturation**—heat the DNA, break the hydrogen bonds, generate two separate strands
      (2) **Annealing**—cool the solution and cause premade primers to bind complementarily to the promotor region of the desired DNA fragment.
      (3) **Replication**—allow heat-stable DNA polymerase to begin replication of the DNA sequence that follows each primer.
   d. These three steps are repeated 20 to 30 times to generate the desired quantity of copies.
   e. As the process proceeds, the numbers of copies grows exponentially! (Because with each cycle, not only is the original fragment being replicated, but all existing copies are being replicated as well.)

Pseudoclinical Correlation

HPI: 42 yo former football star is caught in a high-speed chase while fleeing from law enforcement officials. Samples of blood, hair, and saliva are extracted from the victims' bodies of a murder 15 miles away. The samples are dissolved in various solvents at a federal forensics laboratory. The small amounts of DNA are reconstituted in buffer solution, which is concentrated, digested by DNA enzymes, and amplified by which of the following techniques: PCR: Part II: The RFLP analysis shows a 99.9999% congruence rate of the DNA of these victims and the fleeing football star. Which of the following is most likely the perpetrator of the crime?
   A. The football player
   B. The gunner in your class that purchased this book 8 months before you did
   C. None of the above

(Now, if only the boards were like this.)
3. **Southern blot**
   a. **Goal:** To detect **DNA** fragments that contain a specific base sequence
   b. **Logic:** Have a labeled DNA probe bind complementarily to the DNA fragment of interest
   c. **Steps:**
      (1) Run DNA fragments through gel electrophoresis.
      (2) Transfer ("blot") the DNA fragments onto a filter.
      (3) Immerse the filter in a solution containing a probe that is complementary to the DNA sequence of interest and has a radioactive or fluorescent group. Allow probe to anneal.
      (4) Visualize the DNA fragment of interest using autoradiography or fluorescence.
4. **Northern blot**
   a. **Goal:** To detect **RNA** fragments that contain a specific base sequence
   b. **Logic:** Have a labeled DNA probe bind complementarily to the RNA fragment of interest
   c. **Steps:** Same as with Southern blot, except Northern blot involves radioactive DNA probe binding to a fragment of RNA

Key points:
- What test is used to determine whether a certain gene is expressed?
  → Northern blot (looking for the mRNA of that gene)
- What technique is used to make many copies of a certain section of DNA?
  → PCR

C. **PROCEDURES INVOLVING PROTEINS**
   1. **How do you purify and separate proteins in order to study them?** (Four different methods may be used.)
      a. **Selective precipitation.** Apply pH, heat, or certain ions to a solution so as to cause the protein of interest to react and precipitate out of solution.
      b. **Gel electrophoresis.** Run proteins through a polymer gel matrix that has a positive and a negative end (separates proteins based on size and charge).
      c. **Ion-exchange chromatography.** Use ions to attract proteins (separates proteins based on charge).
      d. **Affinity chromatography.** Add antibodies to a solution so as to bind certain proteins and cause them to precipitate out of solution.
   2. **How do you visualize the three-dimensional structure of a certain protein?**
      → Use NMR (nuclear magnetic resonance) or X-ray crystallography

KEY IDEA: You need to understand these two procedures:
1. **Western blot**
   a. **Goal:** To detect the presence of certain **proteins**
   b. **Logic:** Have labeled antibodies bind to the protein of interest
   c. **Steps:**
      (1) Separate proteins by gel electrophoresis.
      (2) Transfer them to a filter.
      (3) Have radioactive or fluorescent antibodies bind to the protein of interest.
2. **Southwestern blot**
   a. **Goal:** To detect **protein-DNA** interactions (e.g., the binding of transcription factors to the promoter region of a gene).
   b. **Logic:** Expose proteins to a labeled DNA probe.
   c. **Steps:**
      (1) Separate proteins by gel electrophoresis.
      (2) Transfer them to a filter.
      (3) Expose to radioactive or fluorescent DNA fragments.

KEY IDEA: What is recombinant DNA? → Recombinant DNA is human DNA that has been combined with bacterial DNA.

3. **Why would we want to combine human and bacterial DNA?**
   → Two possible reasons:
   a. **To exploit bacteria for the purposes of producing many copies of a certain DNA fragment!** Bacteria multiply rapidly, thus they replicate their genome rapidly. Wouldn't it be great if we could insert a section of human DNA into their genome and exploit their replication rate to generate copies of it? Well, here is how we do it: (Aren't we sneaky?)
      (1) Insert human DNA into bacterial plasmids (by cutting with an endonuclease and then ligating the pieces together using DNA ligase).
      (2) Place the "transformed" bacteria under conditions that optimize their replication.
      (3) Apply the same endonuclease again in order to cut and liberate the desired DNA sections from the bacterial plasmids.
      (4) Isolate and collect the human DNA fragments.
         • **Note:** This process of replicating human DNA by inserting into bacteria cells and culturing those cells is called **"DNA cloning."**
   b. **To exploit bacteria for the purposes of producing human proteins!** Bacteria multiply rapidly and also produce proteins quickly. Wouldn't it be great if we could get them to synthesize human proteins that have important medical uses? Well, here is how we do it:
      (1) Insert human DNA into bacterial plasmids (by cutting with an endonuclease and then ligating the pieces together using DNA ligase).
      (2) Place the "transformed" bacteria under conditions that (a) optimize their replication and (b) optimize production of the desired protein.

(3) Isolate and collect the proteins.
- It's like farming! It's like planting a blackberry bush, watching it spread all over your yard, and then collecting the huge quantities of berries that get produced, which you then put in a bowl and enjoy with vanilla ice cream and pound cake. Mmmmmm. Ahhhhhh. ☺

c. **Clinical correlation** → This is how large quantities of clinically important proteins such as human growth hormone, insulin, and TPA (tissue plasminogen activator) are synthesized.

Chapter 5
PROTEINS

A. PROTEIN STRUCTURE: WHAT YOU NEED TO KNOW

1. An amino acid is composed of three units:
   a. NH_2 or NH_{3+} (amino group)
   b. Alpha carbon with some side chain (R) attached to it
   c. $COOH$ or COO^- (carboxyl group) (Figure 5.1)
2. Under normal conditions (physiologic pH of 7.2), the COOH group will exist as COO^- and the NH_2 group as NH_{3+}. This form is called a **"zwitterion"** (Figure 5.2).
3. **When will the carboxyl group exist in the unionized form (COOH)?** → Only in <u>very acidic conditions</u> (pH of 4 or less), because the pKa for COOH is about 4.
4. **When will the amino group exist in the unionized form (NH_2)?** → Only in <u>very alkaline conditions</u> (pH of 11 or more), because the pKa of NH_{3+} is about 11.

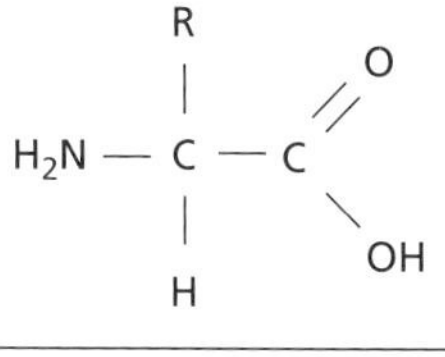

$$H_2N - \underset{\underset{H}{|}}{\overset{\overset{R}{|}}{C}} - C\underset{OH}{\overset{O}{\diagup\diagdown}}$$

FIGURE 5.1 Basic amino acid.

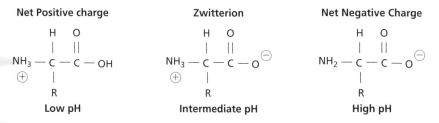

FIGURE 5.2 Relative charges on amino acids.

5. A protein is just a long chain of amino acids linked together by peptide bonds (i.e., "**polypeptide chains**").
6. **Isoelectric point of a protein:** The certain pH at which a protein exists with <u>equal numbers of anions and cations; and thus, no net charge</u>.
7. **How do amino acids link together to form a polypeptide chain?** → The COOH group of one amino acid reacts with the NH3 group of another; losing water in the process and forming a bond called a "**peptide bond**" (Figure 5.3).

<u>**Key idea**</u>: **What are the four important structural characteristics of the peptide bond?**
1. It is **covalent**; therefore it is **strong**.
2. It has **resonance**; therefore it has **partial double bond character** and it is **planar**.
3. It exists in the "**trans**" **configuration**.
4. It is **uncharged, but polar** (because of the oxygens on the COO⁻ group).

---

8. **Are polypeptide chains modified after translation?** → Certainly!! **Three posttranslational modifications:**
   a. Folding
   b. Add sugars, OH groups, or phosphates
   c. Cleave off a portion (sometimes)

B. **The four levels of protein structure—an overview**
   1. **Primary structure:** The order of the amino acids, which are linked together by peptide bonds (see Figure 5.3).
      a. **Bonding involved:** <u>Covalent</u> bonds between amino acids in the <u>main chain</u>.
         (1) These covalent bonds are strong and therefore can only be cleaved by enzymes.

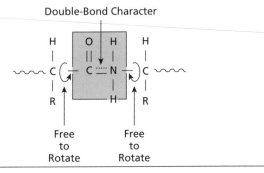

Figure 5.3 Peptide linkage.

> **Clinical correlation → What do covalent bonds in proteins have to do with medicine?**
> **HPI:** 30 yo woman with no history of smoking presents with shortness of breath, especially on exertion. "Not related," but something she wants to share, is her recurrent bladder infections. On PE, lungs appear to have decreased breath sounds bilaterally without any ronchi or rales (bad lung sounds). A barrel-shaped chest is noted as well as hyperresonance on percussion (think of playing the drums). Pulmonary tests show a significant decrease in her FEV compared to her last test 1 year ago (in English, she cannot breathe out as rapidly as she use to). A biopsy is undertaken which confirms emphysema, which is unusual considering her age and nonsmoking status. What is causing her emphysema?
> → Alpha-1-antitrypsin deficiency. (This enzyme deficiency leads to panacinar emphysema.)
> - Proteolytic enzymes are inhibited by alpha-1-antitrypsin. But if alpha-1-antitrypsin is not present, these proteolytic enzymes (e.g., trypsins, elastases, and collagenases) "run wild"—cleaving the covalent bonds of proteins in lung tissue. The result is tissue damage, and eventually, emphysema.

   b. **Secondary structure:** How the chain coils, based on H bonds that develop between main chain amino acids (Figure 5.4).
      (1) Does the chain wind around into a helix (called an "alpha helix"), or does it simply loop back and forth in a plane (called a "beta pleated sheet")?
      (2) **Bonding involved:** H bonds between amino acids in the main chain

KEY IDEA: Because H bonds are fairly weak, the secondary structure CAN be denatured by high temperature, extremes of pH, or strong chemicals.

   c. **Tertiary structure:** How the coiled protein folds when placed in water (see Figure 5.4). **Bonding involved:**
      (1) Free energy change from sequestering hydrophobic amino acids in the middle of the structure—away from water
         exs) leucine, valine, methionine, phenylalanine
      (2) H bonds between polar side chains and water
      (3) H bonds, or ionic bonds, between different side chains
      (4) Disulfide bonds (covalent) between different side chain SH groups

KEY IDEA: If disulfide bonds are present, they lend stability and strength to the tertiary structure (since they are covalent).

   d. **Quaternary structure:** How two or more different folded subunits come together (Figure 5.5).
      ex) Hb subunits come together around a heme group to form an oxygen binding site.
      (1) **Bonding/interactions that may be involved** are the same as those of tertiary structure, EXCEPT there are no covalent disulfide bonds.

KEY IDEA: Because <u>no</u> covalent bonds are involved, the quaternary structure <u>can</u> be denatured by high temperature, extremes of pH, or strong chemicals.

SUMMARY KEY IDEA: All levels of protein structure can be destroyed by high temperature, extremes of pH, or strong chemicals <u>except</u> those that have covalent bonds. What two places are covalent bonds found?
1. Covalent <u>peptide</u> bonds of the primary structure (always present)
2. Covalent <u>disulfide</u> bonds of the tertiary structure (may be present if SH groups are present). These covalent bonds can only be broken by specific enzymes that cleave them. Remember the patient with the alpha-1-antitrypsin deficiency.

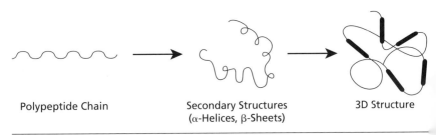

FIGURE 5.4 Primary → secondary → tertiary structures of proteins.

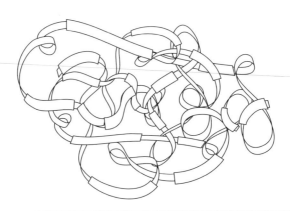

FIGURE 5.5 Quaternary structure of proteins.

C. The four levels of protein structure—detailed explanations of important points.

1. **There are three types of secondary structures:**
   a. **Alpha helix**
      (1) **A single peptide chain** takes the shape of a right-handed twisting helix.
      (2) Bonding involved: **INTRA-chain H bonds** between carbonyl oxygens and amide hydrogens.
      (3) Side chains (aka R groups) extend horizontally outward from the helix (Figure 5.6).
      (4) **Which three amino acids are notorious "alpha-helix breakers?" Why?**
         (a) **Proline**, b/c its NH_2 group cannot hydrogen bond (since it is stuck in a ring) and b/c ring structures are NOT sterically compatible with a spiral helix.
         (b) **Tryptophan**, b/c ring structures are NOT sterically compatible with a spiral helix.
         (c) **Glycine**, b/c it swivels; thus it cannot form a stiff helix.

"What does this have to do with anything?" Check out this patient:

HPI: 7 yo Finnish child presents with chronic malaise. She has habitually returned to clinic complaining of shortness of breath, weakness, pallor, numbness and tingling in the extremities. On PE, glossitis (red tongue) is noted along with hepatosplenomegaly during the abdominal examination. She also has a loss of proprioception and vibratory sense. A CBC shows a macrocytic anemia with hypersegmented PMNs. The diagnosis is confirmed by measuring which vitamin deficiency? → B_{12}.

- This is a vitamin B_{12} deficiency causing pernicious anemia. The lack of B_{12} can be a result of a lack of intrinsic factor, antibodies again parietal cells in the stomach (which secrete intrinsic factor), intestinal malabsorption, or even absence s/p gastrectomy. In this case, the diagnosis is the rare Imerslund-Grasbeck syndrome.
- Normally, vitamin B_{12} binds to intrinsic factor to form a chemical complex the body can use. But this gene mutation alters the **secondary structure** of the protein cubilin that is required to make that complex. (Specifically, a **proline** is mistakenly inserted—screwing up the alpha helix.)

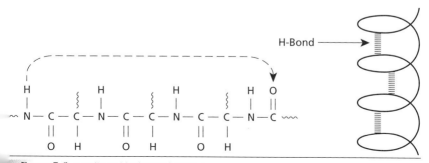

Figure 5.6 α-Helix and hydrogen bonding.

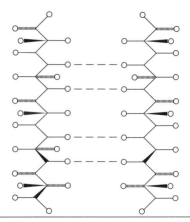

Figure 5.7 Beta sheets.

    b. **B pleated sheet—one-strand version**
       (1) **One single peptide chain** folds back on itself
       (2) Bonding involved: **INTRA-chain H bonds** between carbonyl oxygens and amide hydrogens
       (3) "**Parallel**" configuration (i.e., adjacent sections both run from N to C terminus)
    c. **B pleated sheet—two-strand version**
       (1) **Two separate peptide chains** lay alongside each other
       (2) Bonding involved: **INTER-chain H bonds** between carbonyl oxygens and amide hydrogens lying next to each other
       (3) "**Parallel**" **or** "**antiparallel**" configuration is possible (i.e., the chains next to each other can run in the same direction (both N → C or both C → A), or in opposite directions (one N → C; the other C → N) (Figure 5.7)

D. Protein packaging and distribution
    1. Newly formed proteins are destined for one of 5 possible locations:
       a. <u>Export</u> from the cell
         ex) albumin
       b. Use in the <u>cell membrane</u>
         ex) Na/K pump
       c. <u>Cytoplasm</u>
       d. <u>Mitochondria</u>
       e. <u>Nucleus</u>
    2. **ALL protein translation begins on free-floating ribosomes in the cytoplasm.**
    3. **However, where translation ends depends on the ultimate fate of the protein!**

a. Proteins destined for use within the cell (e.g., cytoplasm, mitochondria, or nucleus)—their translation ends on free-floating cytoplasmic ribosomes.
b. Proteins destined for use in the cell membrane or export out of the cell—their translation ends on the surface of the endoplasmic reticulum.

E. **Collagen: The most abundant protein in the body**
  1. **Synthesis and structure of collagen**
      a. Collagen is the most abundant protein in the body. It is produced in the superficial fascia by cells called "fibroblasts."
      b. The basic unit of collagen is called a **"monomer."** A collagen monomer consists of a triple-stranded helix of three polypeptide chains (like a three-stranded rope). The chains are held together by H bonds. Each of the three chains in the monomer is called an **"alpha chain."**
          (1) There are actually many different kinds of collagen monomers. The differences arise from using alpha chains with slightly different amino acid sequences.
          (2) Regardless of their differences in sequence, all polypeptide alpha chains have glycine as every third amino acid (e.g., X-Y-glycine-X-Y-glycine...).
          (3) The X and Y positions are usually proline, hydroxyproline, or hydroxylysine.
      c. Collagen monomers can be "cross-linked" together with covalent bonds in order to form larger and stronger structures called **"collagen fibrils."**
          (1) This cross-linking occurs between the lysine and hydroxylysine residues.
          (2) **Why is cross-linking of collagen monomers so important?** → B/c it increases the strength of the collagen structure.
      d. **Nutritional correlation: Why does a lack of vitamin C inhibit collagen cross-linking?** → B/c vitamin C is necessary for the synthesis of hydroxyproline and hydroxylysine.
      e. **Which enzyme catalyzes collagen cross-linking?** → Lysyl oxidase
      f. **Nutritional correlation: Why does a copper deficiency inhibit collagen cross-linking?** → B/c copper is required for Lysyl oxidase to function properly.
  2. **Functions of collagen:** There are many different types of collagen, but four of them are most important, and each of the four types predominates in certain areas of the body (Table 5.1).

TABLE 5.1

Type of Collagen	Tissue in Which it Is Found
Type 1	Skin, bone, tendon, blood vessels, cornea
Type 2	Cartilage, intervertebral disk, vitreous body
Type 3	Blood vessels, fetal skin
Type 4	Basement membrane of cells

Clinical correlations
HPI: 67 yo man presents with fatigue, **malaise,** and decreased appetite. The man has lost some weight. His family history is negative for any cancers, and he was in good health until he moved to his new **nursing home** 2 years ago. His only complaint is that his diet is very **bland.** On PE, **hyperkeratosis,** hemorrhagic splinters in the fingers (bloody fingernails), and perifolliculitis (inflammation around the hair follicle) is noted. Furthermore, his gums, nasal, and buccal **mucosa are bleeding.** A CBC shows anemia, and bleeding time is prolonged. What is his problem?
→ Scurvy caused by Vitamin C deficiency
- The three polypeptide chains used to form a collagen monomer each have lots of the molecules **hydroxyproline and hydroxylysine** in them.
- Lack of vitamin C → ↓ hydroxyproline and hydroxylysine synthesis → ↓ H bonding ability between the polypeptide chains → defective collagen monomers → scurvy.

What if it is not a vitamin deficiency, but rather something is messed up to begin with?
HPI: 7 yo boy presents with dislocated right shoulder. He was wrestling with his brother when he heard something "pop." This is the fifth time this year that something like this has happened. The parents adamantly deny any child abuse. On PE, bruises and scars are noted on the knees and elbows. Bilateral sclera have a bluish tint. CBC and Chem 7 are normal. A clotting profile of PT/PTT time is normal as well. The astute medical student makes the diagnosis of…?
→ Ehlers-Danlos syndrome: faulty collagen synthesis.
- Autosomal dominant and recessive mutations in collagen genes result in a variety of forms of Ehlers Danlos.

F. PORPHYRINS: CYCLIC POLYPEPTIDE COMPOUNDS THAT BIND METAL IONS
1. **Pathologies of porphyrins and their related enzyme deficiencies** (Table 5.2)
2. **Porphyrin and heme synthesis** (Figure 5.8)

TABLE 5.2 Porphyrin Pathology and Related Enzyme Deficiencies

Pathology	Deficient Enzyme
Hereditary protoporphyria?	Ferrochetalase deficiency
Acute intermittent porphyria?	Uroporphyrinogen 1 synthetase deficiency
Congenital erythropoietic porphyria?	Uroporphyrinogen 2 cosynthetase deficiency

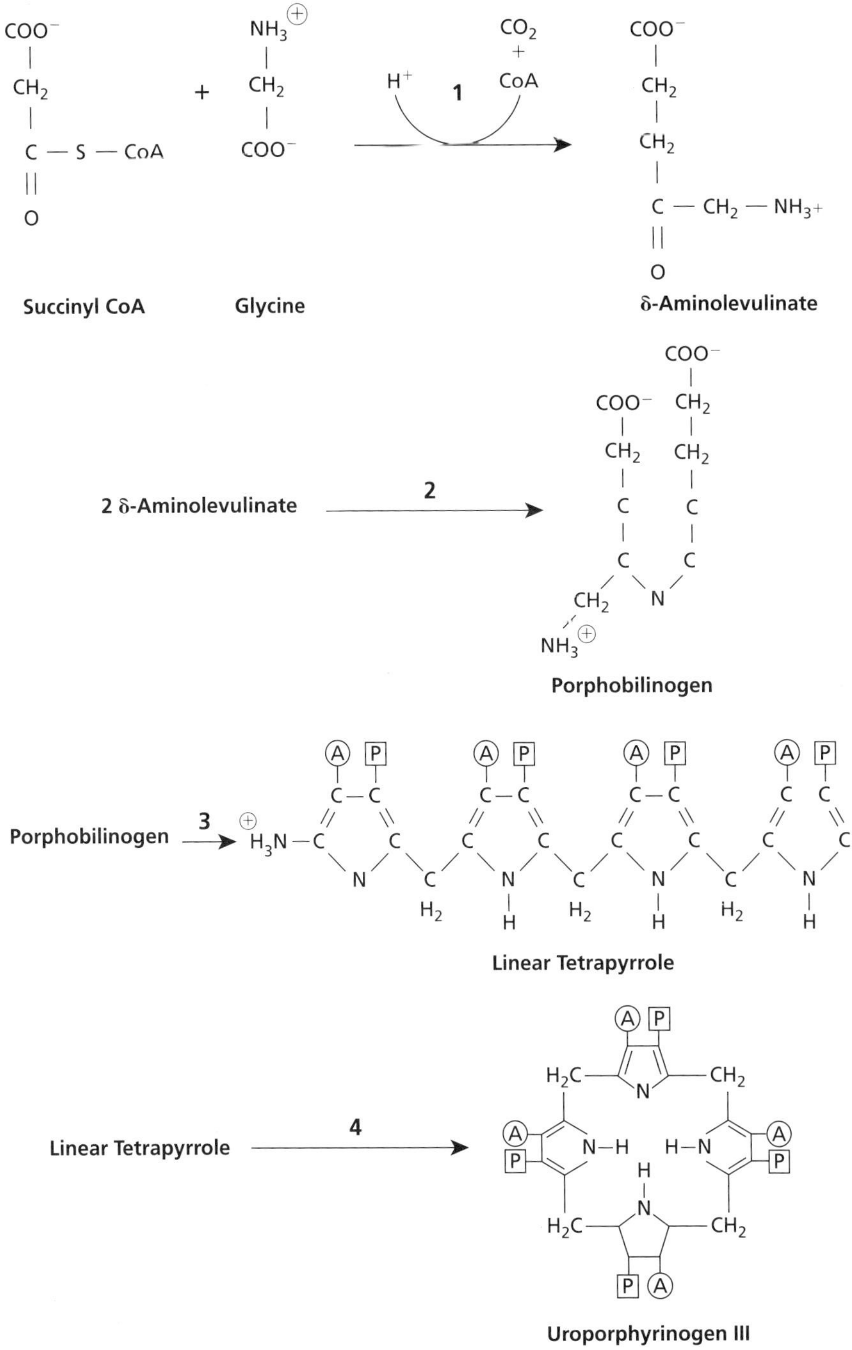

FIGURE 5.8 Porphyrin synthesis and heme synthesis.

Continued

Uroporphyrinogen III —5→ Coproporphyrinogen III

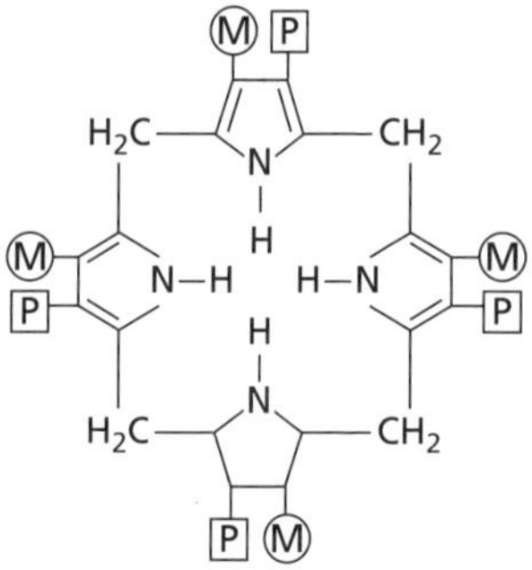

Coproporphyrinogen III —6→ Protoporphyrin IX

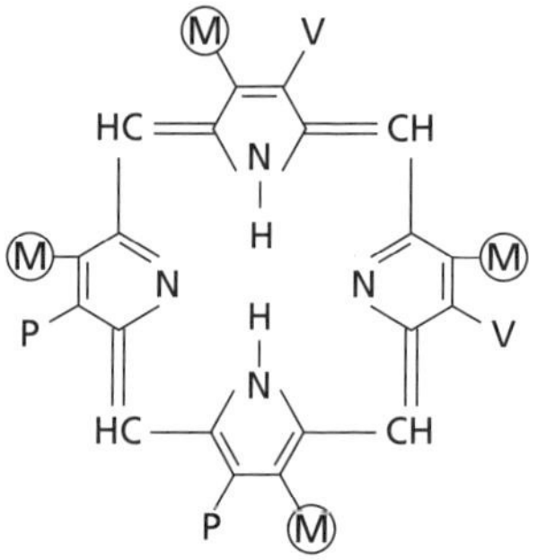

Protoporphyrin IX —7, 8→ Heme

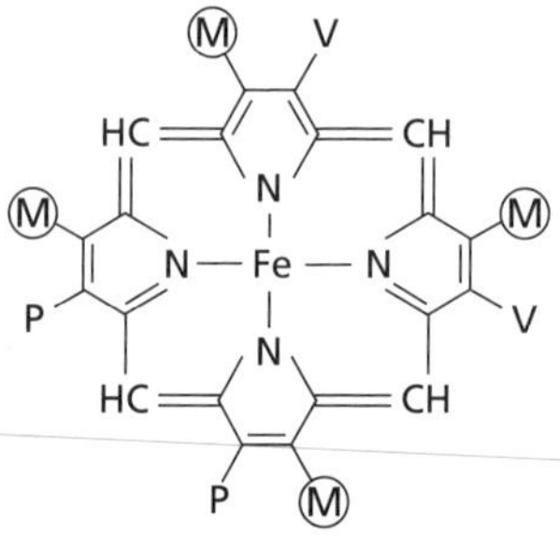

Legend
1. δ-Aminolevulinate synthetase
2. δ-Aminolevulinate dehydrase
3. Uroporphyrinogen I synthetase
4. Uroporphyrinogen III synthetase + uroporphyrinogen III cosynthetase
5. Uroporphyrinogen decarboxylase
6. Coproporphyrinogen oxidase
7. Protoporphyrinogen oxidase
8. Ferrochetalase

(A) Acetate
(P) Proprionate
(M) Methyl
(V) Vinyl

FIGURE 5.8—CONT'D Porphyrin synthesis and heme synthesis.

G. Heme: A porphyrin that binds iron

1. **What is a porphyrin?**
    a. A cyclic polypeptide compound that readily binds metal ions.
    b. One example of a porphyrin is heme (which binds iron).
2. **What are three important roles of heme in the body?**
    a. Structural component of myoglobin
    b. Structural component of hemoglobin
    c. Structural component of the cytochromes of the electron transport chain
3. **Heme synthesis**
    a. **What is the location of heme synthesis?** → Mitochondria AND cytoplasm of the liver and bone marrow.
    b. **What are the two starting reactants of heme synthesis? Where do they come from?**
        (1) **Glycine** (An amino acid. Can come from the diet or from degradation of cellular proteins).
        (2) **Succinyl CoA** (comes from the TCA cycle).
    c. **What are the rate-limiting enzyme and reaction of heme synthesis?**

$$\text{Succinyl CoA + glycine} \xrightarrow{\text{ALA synthetase}} \text{Aminolevulinate (ALA)}$$

    d. **Where does this rate-limiting reaction occur?** → Mitochondria
    e. **What inhibits heme synthesis by inhibiting ALA synthase?** → Newly formed heme (Notice that this is a form of negative feedback.)

Clinical correlations

HPI: 43 yo **black** woman presents to ER with **sharp abdominal pain**. The pain is severe and diffuse. The surgeons **rule out any specific diagnosis** as the patient has had previous exploratory laparotomies (surgical exploration) where no pathology was found. No skin lesions are present. Psychiatry is called to evaluate the patient, as she states that she has had some **hallucinations**. The **great pain** is relieved only with narcotics and the patient is admitted for pain disorder. An astute medical student orders tests for urine porphobilinogen and gamma-aminolevulinic acid, both of which return positive. Are you the astute medical student? What is your diagnosis?
→ AIP: acute intermittent porphyria

- **What are porphyrias?** → Disorders that result from accumulation of intermediate substrates in the heme synthesis pathway
- **What are the two main causes of porphyrias?**
    1. **Inherited deficiencies in certain enzymes**
        - In this case, acute intermittent porphyria is an autosomal dominant lack of porphobilinogen deaminase.
        - What is the other main cause of porphyrias? Try the case on the following page.

Continued

4. **Degradation of heme**
   a. RBCs get old and are absorbed by the spleen and liver for degradation. The heme group gets broken apart into two pieces: The peptide chain and the Fe^{2+} atom. The Fe^{2+} atom is recycled (i.e., used to synthesize new heme), and the polypeptide chain goes on a fantastic voyage in which it eventually becomes bilirubin.
   b. **"Fantastic Voyage of the Polypeptide Heme Chain": It gets converted to bilirubin, which then gets excreted as a part of bile** (Table 5.3) (Figure 5.9).

TABLE 5.3

Site	What Happens
First: Inside a macrophage in the liver or spleen	1. Heme chain gets converted to biliverdin and then to unconjugated bilirubin. heme → biliverdin → bilirubin 2. Unconjugated bilirubin from the spleen gets secreted into the bloodstream.
Second: Bloodstream	Albumin binds unconjugated bilirubin and takes it to the liver. **Why must albumin bind bilirubin?** B/c free, unconjugated bilirubin can cross the BBB and is harmful to the CNS.
Third: Liver	1. Conjugates bilirubin with glucuronate. 2. Releases conjugated bilirubin (bilirubin-diglucuronide) into the bile canaliculi to form part of the bile.
Fourth: Gallbladder	Bile is released into the duodenum.
Fifth: Small intestine	Bacteria hydrolyze bilirubin—diglucuronide into urobilinogen.

Continued

TABLE 5.3—CONT'D

Site	What Happens
Sixth: Terminal ileum of small intestine	Some urobilinogen gets reabsorbed as part of bile reabsorption. It is then sent to the kidneys, where it gets converted into urobilin (a yellow substance) and excreted in the urine. (Urobilin is what makes urine yellow.) **But some urobilinogen does NOT get reabsorbed.**

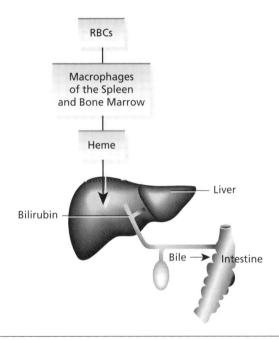

FIGURE 5.9 Heme degradation.

I. QUICK CLINICAL FACTS
1. **Conjugated** bilirubin is called **"direct"** bilirubin.
2. **Unconjugated** bilirubin is called **"indirect"** bilirubin.
3. Total serum bilirubin = direct bilirubin + indirect bilirubin.
4. **What makes urine yellow?** → Urobilin (a breakdown product of bilirubin).
5. **In a jaundiced patient, what makes the eyes and skin yellow?** → Bilirubin in the blood gets converted to urobilin.
6. **What happens to urobilinogen that does not get reabsorbed along with bile in the terminal ileum?** → It enters the large intestine and gets converted by bacteria into stercobilin (a brown substance which gives feces its color). Then it gets excreted as feces.

7. **So, what makes feces brown?** → Stercobilin (a breakdown product of bilirubin).

Now let's hit some cases. What happens if the previous process gets screwed up? Hmmm…

> **HPI:** 2-week-old newborn presents with jaundice that has been worsening since **birth**. The parents state that the yellow complexion began under the tongue, continued from top to bottom. The parents of **Jewish** descent state that this has been a common problem in the family. Blood work shows a markedly increased **unconjugated** bilirubin. What's the diagnosis?
> → Crigler-Najjar syndrome
> - Deficiency of glucuronyl transferase → indirect bilirubinemia → brain damage (if greater than 20 mg/dL)
> - Crigler-Najjar syndrome type I: severe and autosomal recessive
> - Crigler-Najjar syndrome type II: autosomal dominant, less severe

> **HPI:** 18 yo African-American woman presents with "stomach" and right upper quadrant pain for many months. Neither the patient nor her family has any history of gallstones. On PE, **jaundice is noted** on the sclera bilaterally. Blood test reveals slightly elevated LFTs (liver function tests), **increased direct and indirect** bilirubin, and ratio of coproporphyrin I: III of 5:1 (high) in the urine. Ultrasound rules out gallstones or any definable hepatic processes. Make that diagnosis!
> → Dubin-Johnson syndrome: defective bilirubin excretion
> - Benign autosomal recessive

> **HPI:** 35 yo presents with "yellowing" of skin and eyes, especially when he is stressed. He has NO medical history of any problems with his liver (i.e., no IV drug use, hepatitis, transfusions, dark urine, etc.) On PE, a yellow pallor is noted on his complexion, with scleral icterus (yellow eye) and sublingual jaundice (yellow under the tongue). Blood work shows normal LFTs and a bilirubinemia that increases after a fast. What the hey is going on…?
> → Gilbert's disease: unconjugated bilirubinemia secondary to defective uptake by liver cells.

I. **MYOGLOBIN: A RESERVOIR OF OXYGEN FOR CARDIAC AND SKELETAL MUSCLE**
   1. **Important concepts about myoglobin:**
      a. Myoglobin is a structure that consists of a single polypeptide chain (called a globin) surrounding one heme group.
      b. The main feature of the heme group is a Fe^{2+} (iron) atom. This Fe^{2+} atom is capable of reversibly binding one O_2 molecule.

2. **What the heck is metmyoglobin?**
   a. A myoglobin molecule whose Fe atom is 3^+ instead of 2^+
   b. As we will see later, metmyoglobin is <u>bad</u> because Fe^{3+} <u>cannot bind oxygen!</u>

> **KEY IDEA:** Myoglobin has a higher affinity for oxygen than does Hb. Who cares? → This is why heart and skeletal muscles have high concentrations of myoglobin within them. The greater affinity of myoglobin for O_2 is used to draw oxygen away from Hb (i.e., <u>out of</u> the bloodstream) and <u>into</u> the muscle, where it is needed for aerobic respiration.

J. **HEMOGLOBIN (Hb): A PROTEIN THAT BINDS AND TRANSPORTS OXYGEN IN THE BLOOD**
   1. **Hemoglobin structure**
      a. Hemoglobin is a molecule with four polypeptide chains (or globins)—two alpha chains and two beta chains. Each chain contains a heme group (thus an Hb molecule contains four heme groups).
      b. The most important part of each heme group is an Fe^{2+} (iron) atom that is capable of reversibly binding one oxygen molecule.
      c. A red blood cell contains many molecules of hemoglobin within it.
      d. Hb has two forms: T (taut) and R (relaxed). The **T form** has a **low affinity** for O_2. The **R form** has a much **higher affinity** for O_2 (300x greater than the T form).
      e. **What happens when Hb has a <u>LOW affinity</u> for oxygen (i.e., T form)?**
         → There is an advantage and a disadvantage:
         (1) Disadvantage: It is **harder to bind oxygen** from inspired air in the lungs, BUT...
         (2) Advantage: It is **easier to unload oxygen** to metabolically active tissues.
      f. **What happens when Hb has a <u>HIGH affinity</u> for oxygen (i.e., R form)?**
         → There is an advantage and a disadvantage:
         (1) Advantage: It is **easier to bind oxygen** from inspired air in the lungs, BUT...
         (2) Disadvantage: It is **harder to unload oxygen** to metabolically active tissues.
      g. **Summary: There is a tradeoff.** Hb in its low-affinity T form has one main advantage and one main disadvantage. Hb in its high-affinity R form has the exact opposite advantage and disadvantage.
   2. **Cooperative binding and the Hb-O_2 dissociation curve**
      a. Hb exhibits "**cooperative binding.**" This describes the fact that the first oxygen molecule to bind causes a conformational change (from the low-affinity T form <u>to</u> the <u>high</u>-affinity R form) that makes it <u>much easier</u> for all subsequent oxygen molecules to bind. The reverse is also true. Any molecule (such as H^+, CO_2, or DPG) that binds and causes a change from the high affinity R form <u>to</u> the <u>low</u>-affinity T form makes it <u>much harder</u> for subsequent oxygen molecules to bind.

> **KEY IDEA:** What are the four factors that ↓ **the affinity of Hb for oxygen** by changing it to the T form?
> 1. H^+ (acidosis)
> 2. CO_2
> 3. ↑ temperature
> 4. DPG

  b. By ↓ the affinity of Hb for oxygen, these substances <u>facilitate unloading</u> of oxygen <u>to tissues</u>, BUT THEY HAVE THE DRAWBACK OF <u>inhibiting the binding</u> of oxygen <u>from inspired air in the alveoli</u>. This decrease in Hb affinity for oxygen is reflected as an <u>R shift</u> in the Hb-O_2 dissociation curve (Figure 5.10).

  c. Why does it make sense that these four substances would <u>facilitate unloading</u> of oxygen to tissues?

   (1) **Because CO_2, H^+, and ↑ temperature are all byproducts of metabolically active tissues,** and metabolically active tissues have an increased need for oxygen. So essentially, metabolically active tissues in need of O_2 produce substances (CO_2, heat, H^+) that facilitate the delivery of O_2 to these same tissues. Sweet!

   (2) **Because DPG is produced in response to chronic hypoxemia** (e.g., living at high altitudes), and hypoxemic conditions require more oxygen delivery to tissues. So essentially, chronic low O_2 conditions cause ↑ production of a substance (DPG) that facilitates unloading of oxygen to needy tissues.

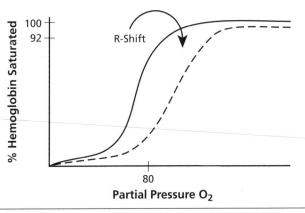

FIGURE 5.10 R-Shift of oxygen dissociation curve.

KEY IDEA: Why is the sigmoid shape of the Hb-oxygen dissociation curve physiologically advantageous? (two reasons)
1. Notice the flat part of the curve at high PO_2 levels. This means that even if the PO_2 in inspired air were to <u>decrease</u> by a sizable amount (i.e., less O_2 were available), <u>most Hb molecules can still be saturated with oxygen!</u>
2. Notice the drop in the curve beginning at PO_2 levels of around 75. Because the PO_2 in tissues ranges from 80 to 40, <u>the drop in the curve over this range represents the sensitivity of Hb to the oxygen needs of tissues</u>. Every decrease in PO_2 that is found among tissues is matched by a willingness of Hb to unload oxygen.

3. Hb's affinity for O_2 versus CO versus CO_2

KEY IDEA: Which does Hb have a greater affinity for? Oxygen, carbon dioxide, or carbon monoxide? → Hb affinity: CO >>> CO_2 > oxygen

KEY IDEA: If Hb has a greater affinity for CO_2 than O_2, then how can Hb <u>release</u> CO_2 and <u>bind</u> O_2 in the lungs? → It is all about the concentration gradients. Inspired air has both <u>more (O_2)</u> and <u>less (CO_2)</u> than does venous blood. Thus oxygen wants to flow down its concentration gradient <u>into</u> the blood, and CO_2 wants to flow down its concentration gradient in the opposite direction—<u>out of</u> the blood and into the lung. Because of these concentration gradients it becomes energetically favorable for Hb to <u>release</u> CO_2 and to <u>bind</u> O_2 in the lungs.

Clinical correlation
HPI: 32 yo man is brought to the emergency room after having left his car on, in his closed garage. He appears to be comatose and is unresponsive. O_2 saturation is 96%. The patient is placed on 10L oxygen on a respirator. Attempts to revive the patient are unsuccessful and the patient expires. A straightforward case of...? → Carbon monoxide poisoning
- **Why is carbon monoxide poisoning so harmful?** → Two reasons:
  1. **The potency of CO**
     a. Small amounts of CO have potent effects, b/c Hb has such a high affinity for CO. (Essentially, every inhaled molecule of CO succeeds in binding to a hemoglobin.)
  2. **The way that CO increases the affinity of Hb for oxygen.**
     a. **CO binds to Hb and changes it to the R form (high affinity for oxygen).** This means that **now Hb has trouble unloading oxygen to tissues that need it!!!**
- **Why is an O_2 sat test NOT helpful in diagnosing carbon monoxide poisoning?** → B/c the problem of carbon monoxide is NOT a lack of Hb being saturated with O_2; it is that Hb is <u>holding onto the O_2 too tightly</u> and cannot release it to needy tissues! B/c it changes Hb to the <u>higher-affinity</u> R form.

Continued

Clinical correlation—cont'd

carbon monoxide actually <u>increases</u> the O_2 sat percentage!! But "el problemo" is that by increasing the affinity of Hb for O_2, the <u>unloading</u> of O_2 to needy tissues is <u>inhibited</u>. How sad—all these hemoglobins filled with oxygen, but they can't release it!!

- **What is the treatment for carbon monoxide poisoning?** → Oxygen therapy. The idea is that by flooding the system with oxygen, you provide enough molecules to out-compete circulating CO for binding sites on the Hb.

K. OXIDIZED FORMS OF HEMOGLOBIN AND MYOGLOBIN

<u>**MONDO KEY IDEA:**</u> If the iron atom (Fe^{2+}) in a heme group gets oxidized to the Fe^{3+} state, this is bad! Very bad!! Muy malo!!! (Why? → B/c Fe^{3+} iron cannot bind oxygen.)

1. When the Fe^{2+} atom in <u>myoglobin</u> is oxidized to become Fe^{3+}, the resultant molecule is called what? → Metmyoglobin
2. When the Fe^{2+} atom in <u>hemoglobin</u> is oxidized to become Fe^{3+}, the resultant molecule is called what? → Methemoglobin
3. What are three things that could cause such oxidations?
   a. Certain drugs
   b. Endogenous free radicals
   c. An inherited mutation
4. **So, why are metmyoglobin and methemoglobin so bad for you?** → B/c they contain Fe^{3+}, which cannot bind oxygen.
5. **What does the body do to correct the problem of metmyoglobin and methemoglobin formation?** → It reduces the Fe^{3+} atom by reacting it with an NADH molecule.

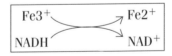

6. **Clinical correlation** → **Why are newborns more harmed by the production of methemoglobin than adults?** → B/c they have lower levels of the enzyme that catalyzes the reaction converting Fe^{3+} → Fe^{2+}.

Chapter 6
Enzymes

A. **How do cells capture and store energy?**
1. Cells need energy to survive. The following important cellular processes all require energy:
    a. Pumping ions against their electrochemical gradients
    b. Forming and moving vesicles
    c. Generating heat
    d. Synthesizing proteins needed to carry out cellular functions
2. **How do cells obtain energy?** → By taking in molecules, breaking their bonds through chemical reactions, and "capturing" the energy released in the process
3. **What chemical reactions does a cell use to generate energy?**
    → The four major catabolic processes of metabolism are:
    a. Glycolysis
    b. Krebs cycle
    c. B oxidation of fatty acids
    d. Deamination of amino acids
4. **How is this captured energy stored by the cell?**
    a. At first, most of it is stored in the bonds of NADH and FADH2.
    b. Ultimately, it is all stored in the high-energy phosphate bonds of ATP or GTP.
5. **How much energy is released from a certain reaction?**
    a. This is represented by the free-energy change (ΔG) of that reaction.
    b. "Free energy" refers to energy that is available to do work.

B. **ΔG: The free-energy change involved in a certain reaction**

> **Key idea:** There are actually two different ΔG measurements. It is important to understand the differences between them, namely, that one is physiologically relevant and the other is not.

1. **ΔG^0 versus ΔG**
    a. **ΔG^0:** Measures the free-energy change of a reaction when there is a 1 M concentration of all reactants and products, and when the pH is 7 (so called "standard conditions"). HOWEVER, this is **NOT physiologically helpful!** Why not? → B/c no reactions in the body are carried out at concentrations of exactly 1 M for all species and at a pH of exactly 7.

b. **ΔG:** Measures the free-energy change of a reaction regardless of pH, and taking into account the concentration of reactants and products that are physiologically present. This is the **physiologically relevant measure!**
   (1) If ΔG is negative, the forward reaction will proceed spontaneously.
   (2) If ΔG is positive, the forward reaction will not proceed spontaneously, but the reverse reaction will, because the ΔG for the reverse reaction is equal in magnitude but opposite in sign to that of the forward reaction.
   (3) The value of ΔG changes as the reaction proceeds. As the reaction approaches equilibrium, ΔG approaches zero. **At equilibrium, ΔG equals zero.**
c. ΔG^0 and ΔG of a reaction are related to each other by the following equation:

$$\Delta G = \Delta G^\circ + RT \ln \frac{(\text{Products})}{(\text{Reactants})}$$

d. If a reaction has a negative ΔG, that means it releases free energy as it proceeds (**"exergonic"**) (Figure 6.1).
e. If a reaction has a positive ΔG, which means it absorbs free energy as it proceeds (**"endergonic"**) (Figure 6.2).

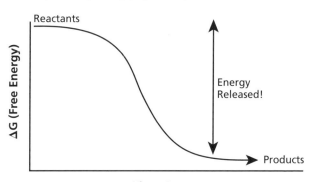

FIGURE 6.1 Exergonic reaction (negative ΔG).

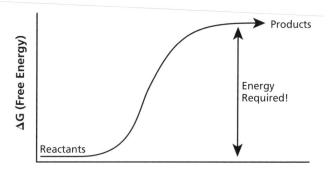

FIGURE 6.2 Endergonic reaction (positive ΔG).

2. **Summary**
   a. **Neg ΔG** = forward reaction **releases free energy** = forward reaction is **spontaneous**; reverse reaction is <u>not</u> spontaneous
   b. **Pos ΔG** = forward reaction **absorbs free energy** = forward reaction is **not spontaneous**; reverse reaction <u>is</u> spontaneous (Figure 6.3)

KEY IDEA: **Let's say the body has an important reaction that it needs to do, but the reaction <u>absorbs</u> free energy (pos ΔG) (thus it does not happen spontaneously). How does the body get this reaction to proceed?** → It has two general strategies:
1. **Change the ΔG of the reaction so that it becomes <u>negative</u>.**
   - How?
       a. By ↑ <u>the amount of reactant</u> present (for example, by increasing the rate of reactant molecule synthesis)
       b. By ↓ <u>the amount of product</u> present (for example, by removing product molecules as soon as they are produced)
       c. By <u>changing the temperature</u>
   - Figure 6.4
2. **Couple this energy <u>absorbing</u> reaction to an energy <u>releasing</u> reaction that is used to "drive it along" ("thermodynamic coupling").**
   a. One common energy-<u>releasing</u> reaction the body uses is the hydrolysis of ATP (Figure 6.5).
   b. Another common energy-<u>releasing</u> reaction the body uses is the flow of ions down their electrochemical gradient.
      ex) The flow of H^+ down its concentration gradient in the mitochondrial intermembrane space is used to drive the hugely energy-absorbing reaction of turning ADP into ATP (Figure 6.6).

Concentration @ Equilibrium	K_{EQ}	ΔG	Type of Reaction
[A] = [B]	1	0	No net movement A ⇌ B
[A] > [B]	<1	>0	Endergonic A ⇌ B
[A] < [B]	>1	<0	Exergonic A ⇌ B

FIGURE 6.3 Spontaneity if determined by ΔG.

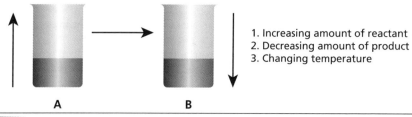

Figure 6.4 Factors changing ΔG.

1. Increasing amount of reactant
2. Decreasing amount of product
3. Changing temperature

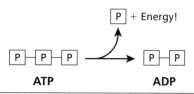

Figure 6.5 Results of hydrolysis.

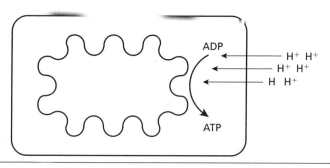

Figure 6.6 Energy from electrochemical gradient.

KEY IDEA: What determines whether a cellular reaction has a positive or a negative ΔG? → A combination of three factors determines the magnitude and sign of ΔG:
1. The direction of heat flow (ΔH)
2. The change in the level of disorder (ΔS)
3. The temperature

---

3. **Definitions**
   a. **Enthalpy (H)** is heat. Therefore ΔH is the amount of heat released or absorbed during a reaction.
   b. **Entropy (S)** is disorder or randomness. Therefore ΔS is the change in disorder or randomness that occurs during a reaction.

c. These factors, plus temperature, determine the delta G for a reaction according to the following relationship:
$$\Delta G = \Delta H - T\Delta S$$
d. **This equation has three important implications:**
   (1) We see that a **negative ΔG is created as a result of a negative ΔH** (heat is released), **and/or a positive ΔS** (disorder increases).
   (2) **At high temperatures, the entropy term (ΔS) becomes more important!**
      ex) At high temps, the fact that a certain amount of heat is released or absorbed (ΔH) does not matter so much as whether disorder ↑ or ↓ (ΔS).
   (3) **At low temperatures, the heat term (ΔH) becomes more important!**
      ex) At low temps, the fact that disorder ↑ or ↓ (ΔS) does not matter so much as whether heat is absorbed or released (ΔH).

C. THE FREE-ENERGY PREDICTION GAME

I tell you some characteristics of the forward reaction. You tell me whether it releases free energy or not (Table 6.1).

D. ENZYMES: PROTEINS THAT INCREASE THE RATE OF A REACTION BY LOWERING THE ACTIVATION ENERGY
1. **Properties of enzymes**
   a. They are not changed or consumed during the course of the reaction.
   b. They ↑ the speed of both the forward <u>and</u> the reverse reaction.
   c. Because they do not change the ΔG, enzymes do not alter the direction or extent of a reaction. They simply affect <u>how fast</u> a reaction occurs.
   d. The magnitude of the increase in reaction rate is extraordinary; a reaction with enzyme catalysis proceeds in the range of 10^3 to 10^8 times as fast as it would uncatalyzed!
   e. Sometimes enzymes need nonprotein molecules (called **"cofactors"**) to be present to catalyze the reaction.

TABLE 6.1 Free-Energy Prediction Game

Does the forward reaction release free energy?	Answer
It releases heat, and it results in greater disorder (neg ΔH) and (neg ΔS)	Yes (regardless of temperature)!
It releases heat, but it creates order (neg ΔH), but (pos ΔS)	It depends on the temperature: • At low temperatures it would release free energy • At high temperatures it would not
It absorbs heat, but it results in greater disorder (pos ΔH), but (neg ΔS)	It depends on the temperature: • At high temperatures it would release free energy • At low temperatures it would not
It absorbs heat, and it creates order (pos ΔH), and (pos ΔS)	No (regardless of temperature)!

(1) One type of cofactor is metal cations (e.g., Zn^{2+}, Fe^{2+})
(2) Another type is small organic molecules (e.g., NAD^+, FAD) (called **"coenzymes"**)

Clinical correlation → **Vitamins** or their derivatives often work as **coenzymes** for crucial reactions.

HPI: 44 yo morbidly obese woman undergoes Roux-en Y gastric bypass procedure to remove a portion of her intestinal tract for weight loss purposes. She returns to clinic 12 months later with shortness of breath, weakness, pallor, numbness and tingling in the extremities. On PE, glossitis (red tongue) is noted along with hepatosplenomegaly during the abdominal exam. She also has a loss of proprioception and vibratory sense. A CBC (blood work) shows a macrocytic anemia with hypersegmented PMNs. The diagnosis is confirmed by measuring:
→ Vitamin B_{12}
- This is a case of B_{12} deficiency leading to pernicious anemia. Remember this example.
- B_{12} deficiency can result from a lack of intrinsic factor, antibodies again parietal cells in the stomach, intestinal malabsorption, or even absence s/p (status post) gastrectomy (removal of the stomach).
- DNA synthesis cannot occur and a megaloblastic anemia develops because B_{12} is needed to maintain tetrahydrofolate, which is needed for DNA synthesis.

---

2. **Mechanism of action of enzymes**
   a. Before reactants can turn into products, they must first pass through an intermediate state called a **"transition state."** The problem with the transition state is that it has a higher energy than reactants do. (This difference between the energy of the reactants and the transition state is called the **"activation energy."**) Thus the reactants face an energy barrier on the path to converting into product.
   b. **From where do reactants get the energy boost needed to form the transition state?** → From their kinetic energy.
   c. **Summary:** Only molecules that can overcome the activation energy barrier can react to form product; and the only way to overcome the activation energy barrier is by kinetic energy. This has three important implications
      (1) At any given time, **only a certain fraction of reactant molecules have sufficient kinetic energy to pass into the transition state.** Thus only a certain fraction reacts to form product per unit time.
      (2) If you could **increase the kinetic energy** of all reactant molecules (e.g., by adding heat), what would happen? More of them per unit time would have sufficient energy to pass into the transition state, and the **reaction rate would increase**.
      (3) If you could **lower the energy of the transition state,** what would happen? More reactant molecules per unit time would find themselves with sufficient kinetic energy to pass into the transition state In other words, the **reaction rate would increase**. This is how enzymes work!!

HUGELY KEY IDEA: How do enzymes increase the rate of a reaction? → By lowering the energy of the transition state, and thereby lowering the activation energy (Figure 6.7).

E. MICHAELIS-MENTEN EQUATION

1. The initial step in an enzyme-catalyzed reaction is the binding of substrate to an active site. Once bound, one of two things can happen: The substrate can unbind (i.e., fall off) or it can react to form product. Each of these steps has a certain rate at which it occurs. These steps can be summarized by the following expression (Figure 6.8):
2. **If enzyme concentration is held constant, and if the enzyme is not allosteric, what happens to the reaction velocity as you increase substrate concentration?** → You get a graph that looks like Figure 6.9:
    a. Notice that as [S] increases, reaction velocity increases, but the rate of increase diminishes over time. Eventually a plateau is reached, at which point continued increases in [S] will <u>not</u> increase reaction velocity any further. Why not? → B/c this is the **point at which all enzymes have become saturated** (i.e., they are all working as fast as they can).
    b. This graph can be described by the Michaelis-Menten equation (Figure 6.10):

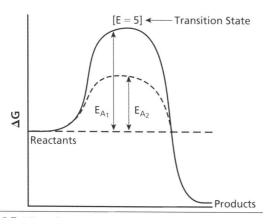

FIGURE 6.7 Effect of enzyme—catalyzed reaction.

$$E + S \underset{k_{-1}}{\overset{k_1}{\rightleftharpoons}} ES \overset{k_2}{\rightleftharpoons} E + P$$

FIGURE 6.8 Steps in an enzyme—catalyzed reaction.

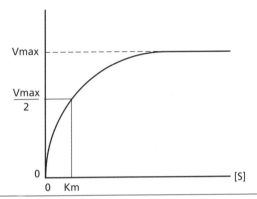

Figure 6.9 Enzyme saturation.

$$V_{initial} = \frac{V_{max}[S]}{K_M + [S]}$$

Figure 6.10 Michaelis-Menten equation

    c. Another way to graph the same information is to present it as a Lineweaver-Burke plot (Figure 6.11), in which the slope = Km / Vmax, the Y intercept = 1/Vmax, and the X intercept = −1/ Km
3. **What is Km and why is it important?**
    a. Km (the Michaelis constant) is **the concentration of substrate needed to produce** a reaction velocity equal to **1/2 Vmax**.
    b. **Km** is important because it indirectly **measures the affinity** of a particular enzyme **for a particular substrate**.
        (1) **Low affinity** of enzyme for substrate = **lots of substrate needed to** bind enough active sites to produce 1/2 Vmax = **high Km**.
        (2) **High affinity** of enzyme for substrate = **little substrate needed** to bind enough active sites to produce 1/2 Vmax = **low Km**.
    c. Because Km reflects an intrinsic property of an enzyme (its affinity for a certain substrate), it does <u>not</u> vary with the concentration of enzyme or substrate.

F. ALLOSTERIC ENZYMES HAVE MULTIPLE ACTIVE SITES THAT INTERACT WITH EACH OTHER
    1. Certain enzymes (called **"allosteric"**) have multiple active sites that interact with each other in such a way that substrate binding at any one site increases the affinity of the remaining active sites for substrate. (This is called **cooperative binding.**)

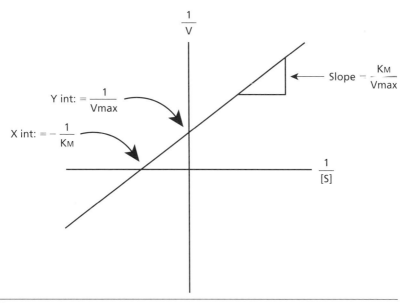

FIGURE 6.11 Lineweaver-Burke plot.

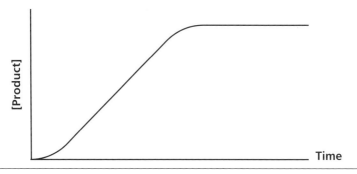

FIGURE 6.12 Sigmoidal curve of allosteric enzyme.

2. **Clinical correlation** → The **hemoglobin** molecule exhibits **cooperative binding of oxygen** between its four different subunits.
3. **Two key points about allosteric enzymes:**
   a. They produce a reaction velocity vs. substrate concentration graph with a **sigmoid shape** (Figure 6.12):
   b. This sigmoid shape is **NOT described by the Michaelis-Menten equation or the Line-Weaver Burke plot.** In other words, allosteric enzymes have nothing to do with the Michaelis-Menten equation or the L-B plot.

> **Why should we care about allostericity?** → **Check this case out:**
> **HPI:** 2 yo child of Mediterranean descent presents with a pale face, feeding difficulties, and general failure to thrive. On PE, the pediatrician notes a mild icterus (yellow eye), maxillary hypertrophy, splenomegaly, and developmental delay on testing. Lab test reveals a microcytic, hypochromic anemia. Target cell are noted on the smear. The diagnosis is...? → Beta-thalassemia
> - Results from decreased synthesis of beta-globin chains. The hemoglobin molecule is composed of both beta and alpha chains. If the beta chains are not present, it is difficult for hemoglobin to form and cooperatively bind oxygen. Allostericity is impaired with this disease.

G. What physiological strategies does the body use to increase the rate of a reaction?

Key idea: Let us say the body needs to <u>increase</u> the rate of a certain reaction. How can it do this?

1. It has four general strategies.
   a. ↑ **The concentration of reactant molecules**
   b. ↑ **The temperature**
   c. ↑ **The quantity of the enzyme** that catalyzes the reaction. **How?**
      (1) ↑ **Transcription** of the gene(s) responsible for making it. The only drawback with this strategy is that it takes a while (hours to days).
      (2) ↓ **The rate of degradation** of the enzyme
   d. ↑ **The speed at which each enzyme works. How?**
      (1) **Have a molecule (a "positive effector") bind at a regulatory site.** This molecule can <u>increase the affinity of the enzyme for substrate, and/or</u> enable the enzyme to <u>lower the activation energy even further</u>.
      (2) **Covalent modification of the enzyme (mainly phosphorylation or dephosphorylation).**
         (a) Some enzymes work fastest when phosphorylated.
         (b) Other enzymes work fastest when dephosphorylated.
      (3) **Change the pH to optimize enzyme function.**
         (a) Each enzyme has a pH at which it works fastest.
         (b) At pH values above or below this optimal pH, enzyme function is compromised. **Why?** → B/c protonation or deprotonation can lower the active site's affinity for substrate and/or cause enzyme denaturation.

Key idea: Let us say the body needs to <u>decrease</u> the rate of a certain reaction. How can it do this? → It can do **the reverse of the steps mentioned in the previous question.** However, we need to take a closer look at how the speed at which an enzyme works can be decreased...

H. How can the function of a single enzyme be inhibited? (i.e., How can the speed at which it works be decreased?)

> **KEY IDEA:** There are two ways in which the function of an enzyme can be inhibited: inhibition at the active site and inhibition at a regulatory site.

1. **Inhibition at the <u>active site</u>** (two types)
   a. **Inhibitor binds to active site <u>reversibly</u>**
      (1) This is called "**competitive inhibition**" (explained further below).
      (2) The inhibitor molecule slows the reaction by occupying the active site for a period of time, thus preventing substrate from binding and reacting.
      (3) **Clinical correlation** → Many medicines that inhibit enzyme function work in this manner. They are **nonreactive substrate analogs**. They compete to bind at the active site, but once bound, they do not react to form product.
         ex) Sulfonamides (an antibacterial) inhibit folate synthesis by acting as nonreactive PABA analogs that compete for the active site of pteroate synthase.
   b. **Inhibitor binds to active site <u>irreversibly</u>.** Enzyme is permanently disabled.

A natural example of enzyme inhibition:

HPI: 35 yo woman in Uganda presents with malaise, nausea, vomiting, and diarrhea. There is no nutritional supplementation of foods where she lives. Additionally, she has been on methotrexate therapy for another illness. On PE, she is pale, but no other abnormalities can be found, including any neurologic dysfunction. A peripheral smear shows a megaloblastic anemia with hypersegmented PMNs. Vitamin B_{12} levels are normal. What is the diagnosis, doctor?
→ Folate deficiency.
- Folate is necessary for DNA methylation and synthesis via pteroate synthase as explained above. Competitive inhibition by methotrexate with folate is thought to be one mechanism by which a megaloblastic anemia arises.

---

2. **Inhibition <u>away from</u> the active site, at a regulatory site** (2 types)
   a. **Inhibitor binds to a regulatory site <u>reversibly</u>**
      (1) This is called "**noncompetitive inhibition**" (explained further below)
      (2) **Clinical correlation** → This is how **Tylenol** (acetaminophen) inhibits cyclooxygenase 1 and 2.
   b. **Inhibitor binds to a regulatory site <u>irreversibly</u>.** Enzyme is permanently disabled.
      (1) **Clinical correlation** → This is how **aspirin** (salicylclic acid) inhibits cyclooxygenase 1 and 2.

I. COMPETITIVE VERSUS NONCOMPETITIVE INHIBITION

1. **Competitive inhibition:** The inhibitor <u>competes with substrate</u> molecules for a chance to bind <u>at the active site</u>. **Two things you need to know about <u>competitive inhibition</u> of a reaction:**
   a. **It does <u>not change</u> the <u>Vmax</u> of the reaction. (Why not?** → B/c it is possible to increase the substrate concentration so much that you overwhelm the inhibitor—never giving it a chance to bind and occupy the active site. At this point, the observed velocity of the reaction is the same as if the inhibitor were never present.)
   b. **It <u>increases</u> the apparent <u>Km</u> of the reaction. (Why?** → B/c Km is the concentration of substrate needed to reach half of the maximal reaction velocity. Because the inhibitor competes with substrate, more substrate than normal is needed in order to occupy the certain number of active sites required to generate 1/2 Vmax.) (Figure 6.13)
2. **Noncompetitive inhibition: the inhibitor molecule binds <u>somewhere other than the active site</u>, thus there is <u>no competition for binding</u>.** Three things you need to know about noncompetitive inhibition of a reaction:
   a. **Vmax of the reaction is <u>decreased</u>. (Why?** → B/c a certain fraction of the enzymes in the reaction are prevented from doing their job. Without these enzymes, the maximum amount of product produced per unit time is lower.)
   b. **This decrease in Vmax can<u>not</u> be overcome by adding more substrate. (Why not?** → B/c substrate binding at the <u>active site</u> has no ability to affect inhibitor binding at <u>a regulatory site</u>.)
   c. **Km of the reaction is <u>unchanged</u>. (Why?** → B/c the binding of inhibitor to a regulatory site does not affect the affinity of the active site for substrate.) **Compare with competitive inhibition in the previous graph** (Figure 6.14).

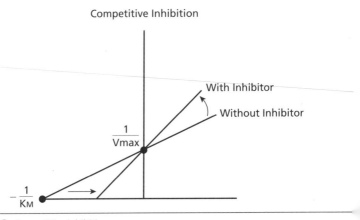

FIGURE 6.13 Competitive inhibition.

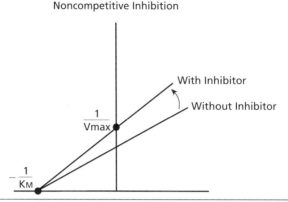

Figure 6.14 Noncompetitive inhibition.

J. Increase in enzyme amount

At times, an increase in enzyme amount or activity is required. Simple modulation is not sufficient. An increase from the genetic level (i.e., an increase in synthesis via transcription and translation) is required.

> **Clinical correlation**
> **HPI:** 23 yo Asian man presents with a severely erythematous complexion. He appears slightly stuporous and possibly intoxicated, although his friends insist he had only one beer. On further questioning, they concede they have no knowledge of the patient's tolerance. The patient is hydrated and recovers from his apparent inebriation. A slightly ruddy complexion remains but much less than his clownlike appearance on admission. The diagnosis is...?
> → Asian male syndrome from cytochrome p450 lack of alcohol dehydrogenase.
> - This is a prime example of the need for upregulation of cytochrome p450. Upregulation of p450 increases tolerance but also liver cirrhosis.

Chapter 7
Cellular Second-Messenger Systems

A. **Interactions between Extracellular Molecules and the Intracellular Environment**

The ways in which an extracellular molecule can produce effects within a cell can be grouped as follows:
1. **Enter the cell, bind to an intracellular receptor in the cytoplasm, travel to nucleus, stimulate transcription of a specific gene**
    ex) Steroid hormones (e.g., testosterone, estrogen), vitamin D, retinoic acid, thyroxine)
2. **Enter the cell and stimulate nitric oxide synthase, which activates the following cascade: NO → guanylate cyclase → cGMP → protein kinase G**
    ex) Organic nitrate induced smooth muscle relaxation, the Viagra thing.
3. **Bind to a <u>receptor</u> on the <u>cell membrane</u> surface, which then causes...**
    a. **An ion channel to open**
       ex) GABA receptors, cholinergic nicotinic receptors
    b. **Activation of a G protein, which goes and opens or shuts ion channels**
    c. **Activation of the intracellular enzymatic portion of the receptor**
       ex) Insulin induced activation of intracellular tyrosine kinase
    d. **Activation or inhibition of the following cascade: G protein → adenylate cyclase → cAMP → protein kinase A**
       ex) Glucagon, epinephrine, alpha and beta adrenoreceptors
    e. **Activation of the following cascade: G protein → guanylate cyclase → cGMP → protein kinase G**
       ex) Visual signal transduction
    f. **Activation of the following cascade: G protein → phospholipase C → DAG, IP3 → Ca → protein kinase C**
       ex) Human growth hormone, alpha 1 adrenoreceptors

Who cares about second messenger systems? Check out this case:
HPI: 4 yo child presents with leg cramps and constipation since birth. The child is at the 5th percentile for weight and height. A PE reveals a "Charlie Brown" physique with a round face and short neck. The patient has a positive Chvostek (tapping on cheek causes facial twitch) and Trousseau (blood pressure cuff to cut off arterial circulation causes contractions) signs. Hypocalcemia is diagnosed. Blood work shows normal PTH (parathyroid hormone) levels. What is your diagnosis? What is the secondary messenger defect? If you already know the answer, why are you reading this book? Only an endocrinology attending should know the answer. ☺
→ Seabright-Bantam syndrome (aka pseudohypoparathyroidism).
- PTH resistance on the renal tubule via a defect in response of the **cAMP second messenger system**; an X-linked dominant disorder.

B. Overview: The five most physiologically important second-messenger systems

Key idea: What are second-messenger systems? Why are they important? → They initiate intracellular activities in response to binding of extracellular ligands; their purpose is to transmit and amplify signals from peptide hormones or neurotransmitters (which cannot enter the cell).

Key idea: Here are the five main second-messenger systems in the body. Know them. Love them.
1. G protein linked to **adenylate cyclase**
   - G protein → adenylate cyclase → cAMP → protein kinase A cascade
     ex) Glucagon, epinephrine, alpha and beta adrenoreceptors
2. G protein linked to **membrane-bound guanylate cyclase**
   - G protein → membrane-bound guanylate cyclase → cGMP → protein kinase G
3. G protein linked to **phospholipase C**
   - G protein → phospholipase C → DAG, IP3 → ↑ Ca → ↑ protein kinase C activity
     ex) Human growth hormone, alpha 1 adrenoreceptors
4. Nitric oxide stimulation of free-floating, **cytosolic guanylate cyclase**
   - NO → cytosolic guanylate cyclase → cGMP → protein kinase G
     ex) Vascular smooth muscle relaxation
5. Cell membrane receptor with an **intracellular tyrosine kinase domain**
   ex) Insulin receptors

C. **Detailed summary: The five most physiologically important second-messenger systems**

Remember what their job is → To <u>transmit</u> and <u>amplify</u> signals from extracellular ligands (usually peptide hormones or neurotransmitters) (Figure 7.1) (Tables 7.1 and 7.2)

D. **Summary: Comparing and contrasting the three types of G protein-linked second-messenger cascades**
1. <u>Level 1: Adenylate cyclase vs. guanylate cyclase vs. phospholipase C</u> (Table 7.3)
2. <u>Level 2: cAMP vs. cGMP vs. IP3 and DAG</u> (Table 7.4)
3. <u>Level 3: Protein kinase A versus protein kinase G versus protein kinase C</u> (Table 7.5)

E. **Malfunctions of G protein–linked second-messenger systems**
1. Let us say there is <u>not enough</u> intracellular activity. What could be going wrong?
   a. The <u>ligand is not binding properly</u> to the receptor.
   b. The <u>intracellular conformational change</u> of the receptor (which normally opens up the G protein binding site) <u>is not happening</u>.
   c. The <u>G alpha subunit cannot bind</u> to the receptor.
   d. The <u>G alpha subunit cannot leave</u> the receptor (bound too tightly).
2. Let us say there is <u>too much</u> intracellular activity. What could be going wrong?
   a. The conformational change in the receptor is happening by itself (without a ligand binding), thus opening a binding site for G alpha subunit activation. This is called a **"constitutively active receptor."**
   b. Intrinsic GTPase activity of the G alpha subunit fails to work thus causing the G alpha subunit to stay activated indefinitely.

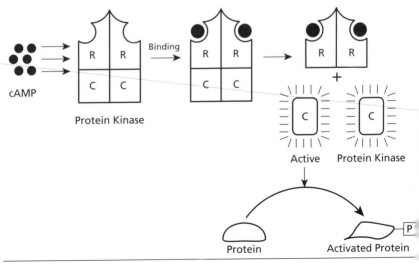

Figure 7.1 What does cAMP do?

TABLE 7.1 Pathways of Important Second Messenger Systems

	G protein linked to cAMP		G protein linked to cGMP	Nitric oxide linked to cGMP
How to turn system on?	Ligand binds receptor in cell membrane		Ligand binds receptor in cell membrane	NO is synthesized by NO synthase
Cascade of events?	↓ Opens Gs (or Gi) protein binding site ↓ Gs (or Gi) protein slides over and binds ↓ Gs (or Gi) alpha subunit drops GDP, grabs GTP, breaks away from beta and gamma subunits, slides toward adenylate cyclases adenylate cyclases ↓		↓ Opens G protein binding site ↓ G protein slides over and binds ↓ G alpha subunit drops GDP, grabs GTP, breaks away from beta and gamma subunits ↓ G alpha subunits stimulate adenylate cyclases ↓ ↑ cAMP ↓ ↑ activation of protein kinase G ↓ ↑ phosphorylation of cellular enzymes or ion channels	↓ NO stimulates adenylate cyclases ↓ ↑ cAMP ↓ ↑ activation of protein kinase G ↓ ↑ phosphorylation of cellular enzymes or ion channels
	Gs alpha subunits stimulate adenylate cyclases ↓ ↑ cAMP ↓ ↑ activation of protein kinase A ↓ ↑ phosphorylation of cellular enzymes or ion channels	Gi alpha subunits inhibit ↓ ↓ cAMP ↓ ↓ activation of protein kinase A ↓ ↓ phosphorylation of cellular enzymes or ion channels		
How to turn system off at the source?	• Ligand dissociates • Intrinsic GTPase activity of alpha subunit ("built-in timer") (i.e., after a period of time the GTPase hydrolyzes GTP, thereby stopping activity of the G alpha subunit) • Endocytosis of ligand/receptor complex		• Ligand dissociates • Intrinsic GTPase activity of alpha subunit ("built-in timer") (i.e., after a period of time the GTPase hydrolyzes GTP, thereby stopping activity of the G alpha subunit) • Endocytosis of ligand/receptor complex	Inactivation of NO synthase

TABLE 7.2 Pathways of Important Second Messenger Systems

	G Protein Linked to IP3 and DAG	Receptor with Intracellular Tyrosine Kinase Activity
How to turn system on?	Ligand binds receptor in cell membrane	Ligand binds receptor in cell membrane
Cascade of events?	↓ Opens Gq protein binding site ↓ Gq protein slides over and binds ↓ Gq alpha subunit drops GDP, grabs GTP, breaks away from beta and gamma subunits, slides toward phospholipase C ↓ Gq alpha subunits stimulate phospholipase C ↓ ↑ IP3 ...and... ↑ DAG ↓ ↓ Release of Ca — Helps protein kinase C bind Ca-calmodulin from sarcoplasmic reticulum ↓ ↓ ↑ Ca-calmodulin complex formation — ↑ activity of protein kinase C ↑ phosphorylation of cellular enzymes	↓ Extracellular and intracellular domains dimerize ↓ Intracellular tyrosine kinase domains phosphorylate each other, thus becoming active ↓ Conformational changes open up active sites for certain cellular enzymes ↓ Cellular enzymes bind and are phosphorylated
How to turn off system at the source?	• Ligand dissociates • Intrinsic GTPase activity of alpha subunit ("built-in timer") (i.e., after a period of time the GTPase hydrolyzes GTP, thereby stopping activity of the G alpha subunit) • Endocytosis of ligand and receptor complex	• Ligand dissociates • Endocytosis of ligand and receptor • Membrane-bound phosphotase inactivates the tyrosine kinase activity and closes the active site

TABLE 7.3 Comparing and Contrasting G Protein–Linked Systems: Level 1

	Adenylate Cyclase	Bound Membrane-Guanylate Cyclase	Free-Floating, Cytosolic Guanylate Cyclase	Phospholipase C
What does it do?	Catalyzes cAMP synthesis from AMP	Catalyzes cGMP synthesis from GMP	Catalyzes cGMP synthesis from GMP	Catalyzes IP3 (inositol triphosphate) and DAG (diacylglycerol) synthesis from PIP2 (phosphatidylinositol)

Continued

CELLULAR SECOND-MESSENGER SYSTEMS

TABLE 7.3 Comparing and Contrasting G Protein–Linked Systems: Level 1—CONT'D

	Adenylate Cyclase	Bound Membrane-Guanylate Cyclase	Free-Floating, Cytosolic Guanylate Cyclase	Phospholipase C
How is it turned on?	Stimulation by Gs alpha subunit	Stimulation by a G alpha subunit	Stimulation by NO	Stimulation by a G alpha subunit
How is it turned off?	1. Inactivation of Gs alpha subunit; 2. Stimulation by Gi subunit	Inactivation of G alpha subunit	Degradation of NO	Inactivation of G alpha subunit

TABLE 7.4 Comparing and Contrasting G Protein–Linked Systems: Level 2

	cAMP	cGMP	IP3 and DAG
Synthesized from	AMP	GMP	PIP2 (phosphatidylinositol) a phospholipid in the cell membrane
Synthesized by	Adenylate cyclase	Guanylate cyclase	Phospholipase C
What does it do?	Activates protein kinase A	Activates protein kinase G	• IP3 causes ↑ Ca release from sarcoplasmic reticulum • DAG helps protein kinase C bind Ca-calmodulin complex (and Ca-calmodulin complex increases the activity of protein kinase C)
How is it inactivated?	By phosphodiesterase	By phosphodiesterase	IP3 is inactivated by a specific phosphotase DAG is metabolized by ?

TABLE 7.5 Comparing and Contrasting G Protein–Linked Systems: Level 3

	Protein kinase A	Protein kinase G	Protein kinase C
Activated by	cAMP	cGMP	DAG
What does it do?	Phosphorylates ion channels and intracellular enzymes	Phosphorylates ion channels and intracellular enzymes	Phosphorylates ion channels and intracellular enzymes
How is it inactivated?	By a protein phosphotase	By a protein phosphotase	By a protein phosphotase

Let us take a G-protein–mediated clinical example:
HPI: 14 yo who has a long-standing history of **bone disease,** with fractures, asymmetry and deformity of the legs, arms, and skull; **endocrine disease,** including early puberty with menstrual bleeding; and early development of breasts and pubic hair and an increased rate of growth. She also has had **skin changes,** with areas of increased pigment distributed in an asymmetric and irregular pattern. She has been treated separately by an orthopedic surgeon, endocrinologist, and dermatologist. She has never received a diagnosis, but being the astute medical student that you are, you realize this triad is a G-protein mediated disorder termed:
→ Mc-Cune Albright syndrome.
- Pituitary hyperfunction including hyperthyroidism, excessive growth hormone secretion, and sexual hormone secretion; associated bone disease (polyostotic fibrous dysplasia) and skin pigmentation (cafe-au-lait spots) are also thought to be endocrinologically mediated. The underlying disorder which research this year has elucidated, is a defect in the GNAS1 gene leading to G-protein overactivation and excess cAMP production.

CHAPTER 8
OVERVIEW OF METABOLISM WITH EMPHASIS ON HORMONAL CONTROL

A. GENERAL PRINCIPLES
   1. **Absorptive versus postabsorptive period** (Table 8.1)
   2. **Important metabolic reactions that are "opposites"** (Table 8.2)
   3. **An analogy we must understand:** Putting molecules through catabolic processes to produce ATP is similar to putting logs through a fire to produce energy.

TABLE 8.1 Absorptive versus Postabsorptive State

Absorptive Period	Postabsorptive Period
Fuel is immediately available in blood (up to 4 hrs after a meal)	Fuel is not immediately available in the blood
Main fuel = circulating glucose from meal	Main fuels = glucose from glycogen, fat, amino acids from breakdown of muscle proteins
• Excess fuel molecules are stored • Excess fuel molecules are used to synthesize important physiologic materials ("anabolism")	• Fuel molecules are released from storage • Physiologic materials are degraded to be used as fuel ("catabolism")
Characterized by insulin secretion	Characterized by glucagon secretion

TABLE 8.2 Two Opposite Types of Reactions: Catabolic versus Anabolic

Catabolic	Anabolic
Glycolysis (glucose → pyruvate)	Gluconeogenesis (pyruvate → glucose)
Glycogen breakdown (glycogen → glucose-6-P)	Glycogen synthesis (glucose-6-P → glycogen)
Triglyceride breakdown (TG → glycerol + 3 FFAs)	Triglyceride synthesis (glycerol-P + 3 FFAs → TG)
Beta oxidation (1 FFA → manyAcetyl CoAs)	De Novo FFA synthesis (many Acetyl CoAs → 1 FFA)

a. The fuel source = "the logs" = the food molecules (amino acids, carbohydrates, fats, ethanol)
b. The site of oxidation = "the fire" = the catabolic processes (glycolysis, TCA, beta oxidation)
c. The energy produced = "heat and light" = ATP
4. **What are the only five sources of energy for the body?**
   a. Carbohydrates
   b. Fats
   c. Proteins
   d. Ethanol
   e. Nucleotides (used only in dire emergencies)
   f. **Which of these substrates provides the most energy per unit weight?**
      (1) Fats (9 Kcal/g)
      (2) Ethanol (7 Kcal/g)
      (3) Carbohydrates and protein (4 Kcal/g)
   g. **When each of these substrates is used for energy, which catabolic pathway does it enter?**
      (1) **Carbohydrates**—broken into monosaccharides; monosaccharides enter **glycolysis**
         - Glucose enters at the beginning of glycolysis
         - Fructose and galactose enter at the middle of glycolysis
      (2) **Fats** (mostly triglycerides)—broken down into glycerol-1-phosphate and three free fatty acids.
         - **Glycerol phosphate** enters **glycolysis**
         - **Fatty acids** are first converted to acetyl CoA or propionyl CoA, and then enter the **TCA cycle**
      (3) **Proteins**—broken down into amino acids. All amino acids can be deaminated to yield carbon skeletons that enter the **TCA cycle**.
         - Some amino acids yield a carbon skeleton of pyruvate.
         - Some amino acids yield a carbon skeleton of acetyl CoA.
         - Some amino acids yield carbon skeletons that are TCA cycle intermediates.
      (4) **Ethanol**—converted into acetyl CoA, which enters the **TCA cycle**.
      (5) **Nucleotides**—DNA and RNA can both, ultimately, be broken down to yield ribose. Ribose can enter **glycolysis** (after first flowing through the HMP shunt).
5. **Under low blood glucose conditions, which of the five fuel substrates can be used to create glucose through a series of reactions called "gluconeogenesis?"** → All of them except ethanol
   a. Carbohydrates—the monosaccharides galactose and fructose can enter gluconeogenesis.
   b. Fats—the glycerol phosphate formed from the breakdown of a triglyceride can enter gluconeogenesis.
   c. Proteins—all amino acids except leucine and lysine are capable of producing carbon skeletons that can enter gluconeogenesis
   d. Nucleotides—the 5-carbon sugar (ribose or deoxyribose) is capable of flowing through the hexose monophosphate shunt and entering gluconeogenesis

6. **The two main mechanisms the body has for ↑ blood glucose:**
   a. **Gluconeogenesis.** (How does it ↑ blood glucose? → The liver and kidneys take **different substrates and turn them into glucose**, which is then released into the blood.)
   b. **Glycogenolysis. (How does it ↑ blood glucose?** → Liver and muscle cells break down glycogen into glucose (using the enzyme glycogen phosphorylase). **Glucose from liver glycogen can be released into the blood** for use by other tissues. Glucose from muscle glycogen CANNOT (i.e., it can only be used intracellularly). Hormones that ↑ blood glucose will often <u>stimulate</u> glycogen <u>breakdown</u> and simultaneously <u>inhibit</u> glycogen <u>synthesis</u>.
7. **The two main mechanisms the body has to ↓ blood glucose levels:**
   a. **Glycolysis. (How does it ↓ blood glucose?** → **By cellular uptake of glucose** from the blood. This glucose is then turned into pyruvate [generating NADH and ATP in the process].)
   b. **Glycogen synthesis. How does it ↓ blood glucose?** → **Liver and muscle cells uptake glucose** from the blood and use it to synthesize glycogen (using the enzyme glycogen synthase).
8. **Metabolism has four goals that are accomplished in different ways:**
   a. **Oxidation of food molecules to produce NADH, $FADH_2$, or ATP.** How is this accomplished?
      (1) Glycolysis
      (2) TCA cycle
      (3) Beta oxidation of FFA
   b. **Synthesis of needed chemicals and materials from food molecules.** How is this accomplished?
      (1) Glycogen synthesis
      (2) Protein synthesis
      (3) FFA and TG synthesis
      (4) Purine/pyrimidine synthesis
      (5) HMP shunt (creates substrates for purine and pyrimidine synthesis; also creates NADPH)
      (6) Cholesterol and steroid hormone synthesis
   c. **Maintenance of a constant supply of fuel for cells to use in generating ATP.** How is this accomplished?
      (1) Breakdown of glycogen, TGs, and proteins (i.e., breakdown of stored fuel supplies)
      (2) Gluconeogenesis
      (3) Cori cycle (muscles send lactate to the liver as a substrate for gluconeogenesis)
      (4) Glucose/alanine cycle (muscles send alanine to the liver as a substrate for gluconeogenesis)
      (5) Ketone formation
   d. **Detoxification of injurious compounds (especially NH_{4+} and unconjugated bilirubin—both of which are toxic to the CNS).** How is this accomplished?
      (1) Urea cycle (converts NH_{4+} into urea for excretion via the urine)

(2) Glucose/alanine cycle (alanine carries NH_{4+} from the muscles to the liver for entry into the urea cycle)
(3) Glutamate/glutamine cycle (glutamine carries NH_{4+} from all peripheral tissues to the liver or kidney)
(4) Conjugation of bilirubin in the liver (converts unconjugated bilirubin into conjugated —which is not toxic to the CNS)
(5) P450 enzyme action in the liver

KEY IDEA: **The body has two strategies for making ATP:**
1. **Oxidative phosphorylation:** Create NADH and/or $FADH_2$, send them to the electron transport chain, generate an H^+ gradient, use the H^+ gradient to drive ATP synthase. The following reactions all participate in oxidative phosphorylation by creating NADH or $FADH_2$ (Table 8.3):
2. **Substrate level phosphorylation:** Take a phosphate and transfer it directly onto an ADP. The following reactions are all examples of substrate level phosphorylations (Table 8.4).

9. The two hormones that <u>most strongly</u> affect metabolism are:
   a. Insulin from pancreas beta cells.
   b. Glucagon from pancreas alpha cells.

BIG DADDY KEY IDEA: Metabolic hormones and their intracellular effects
1. **Insulin 5 tyrosine kinase receptor** = ↓ (cAMP) = DEphosphorylates enzymes
2. **Glucagon, cortisol, epinephrine** = G-protein linked receptor = ↑ (cAMP) = ↑ protein kinase A = phosphorylates enzymes

TABLE 8.3 Reactions Involved in Oxidative Phosphorylation

Oxidative Reaction	Location	Main Substrates	Main Products
Glycolysis	Cytoplasm of all cells	Glucose	2 pyruvate 2 ATP 2 NADH (if aerobic conditions)
Decarboxylation of pyruvate	Mitochondrial matrix of all cells	Pyruvate	1 NADH 1 CO_2
TCA cycle	Mitochondrial matrix of all cells	Acetyl CoA	(per acetyl CoA) 3 NADH 1 $FADH_2$ 1 GTP 2 CO_2
Beta oxidation of FFAs	Mitochondrial matrix of all cells, except RBCs and brain	Free fatty acids	8 acetyl CoA 7 NADH 7 NADH
Oxidative deamination of glutamate	Mitochondrial matrix of all cells, but primarily liver and kidney	Glutamate	1 alpha ketoglutarate 1 NADH (or occasionally 1 NADPH)

TABLE 8.4 Reactions Involved in Substrate Level Phosphorylation

Substrate Level Phosphorylation Reaction	Location	Main Substrates	Main Products
Glycolysis	Cytoplasm of all cells	Glucose	2 Pyruvate 2 ATP (both from substrate level phosphorylations) (if aerobic conditions—also 2 NADH)
ATP synthase	Mitochondrial inner membrane of all cells	H^+ and ADP	ATP
Creatine kinase	Cytoplasm of skeletal muscle	Creatine 1 phosphate and ADP	Creatine molecule 1 ATP

B. Catabolism: Breaking Down Substrates for Energy

1. **Catabolic processes categorized by fuel substrate and location** (Table 8.5)
2. **Fuel catabolism in major metabolic tissues** (Table 8.6)

Key Idea: The following three tissues CANNOT use plasma FFAs for fuel.
1. Brain
2. RBCs
3. Adrenal medulla

3. **A close-up look at metabolism in an exercising muscle**
   a. When muscles are working hard, doing **anaerobic exercise...**
   b. **Where does their glucose come from?**
      (1) Stored muscle glycogen
      (2) Blood glucose
   c. **How can they get more glucose?** → Send alanine or lactate to the liver for gluconeogenesis.
   d. **How can they regenerate NAD^+ so that glycolysis can continue?**
      → Convert pyruvate to lactate

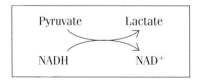

Table 8.5 Fuel Substrates and Catabolic Processes They Participate In

Fuel Substrate	Catabolic Process	Location	Description
Carbohydrates	• Conversion of poly and disaccharides into monosaccharides	• GI tract • Cytoplasm of liver	• Polysaccharides → glucose, fructose, or galactose
	• Glycogen breakdown	• Cytoplasm of liver and skeletal muscle	• Glycogen → G-6-P
	• Entry of monosaccharides or G-6-P into glycolysis	• Cytoplasm of all cells	• Glucose, fructose, galactose, or G-6-P → pyruvate + energy
	• Decarboxylation of pyruvate into acetyl CoA	• Mitochondrial matrix of all cells	• Pyruvate → acetyl CoA + energy
	• Entry of acetyl CoA into TCA cycle	• Mitochondrial matrix of all cells	• Acetyl CoA + OAA → OAA + energy
Fats	• TG breakdown	• Cytoplasm of liver and adipose cells	• TG → free fatty acids + glycerol
	• Conversion of glycerol to glycerol-phosphate	• Cytoplasm of liver	• Glycerol + ATP → glycerol-phosphate
	• Entry of glycerol phosphate into glycolysis	• Cytoplasm of all cells	• Glycerol phosphate → pyruvate + energy
	• Decarboxylation of pyruvate into acetyl CoA	• Mitochondrial matrix of all cells	• Pyruvate → acetyl CoA + energy
	• Beta oxidation of free fatty acids into acetyl CoA	• Mitochondrial matrix of all cells except the brain and RBCs	• Free fatty acids → acetyl CoA
	• Entry of Acetyl CoA into TCA cycle	• Mitochondrial matrix of all cells	• Acetyl CoA + OAA OAA + energy
Proteins	• Protein breakdown	• Cytoplasm of all cells; but mainly liver and skeletal muscle	• Protein → amino acids
	• Transamination of amino acids to yield carbon skeletons	• Mitochondrial matrix of all cells; but mainly liver and skeletal muscle	• Amino acids → carbon skeletons
	• Entry of carbon skeletons into TCA cycle	• Mitochondrial matrix of all cells	• Carbon skeletons → TCA cycle → energy

e. **What are the two sources of ATP for exercising muscles?**
   (1) Glycolysis (two ATP created per molecule of glucose)
   (2) Creatine kinase reaction (one ATP per molecule of creatine phosphate)

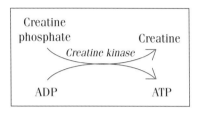

TABLE 8.6 Fuel Catabolism in Major Tissues

Tissue	Energy Sources in Order of Preference	Stored Fuels	Fuel Substrates Exported into Circulation
CNS	• Glucose • Ketones (Brain cannot use plasma FFA or amino acids b/c they have trouble crossing the BBB)	None	None
Liver	• Glucose • Fatty acids (Liver cannot use ketones for energy b/c it lacks the necessary enzymes)	• Glycogen • TG	• Glucose • TG (exported in VLDLs) • Free fatty acids • Glycerol • Ketone bodies
Skeletal muscle (under exertion)	• Glucose • Fatty acids • Ketones	• Glycogen • Creatine phosphate • Muscle proteins	• Lactate (goes to the liver for gluconeogenesis; part of Cori cycle) • Alanine (goes to the liver for gluconeogenesis; part of glucose/alanine cycle)
Skeletal muscle (resting)	• Fatty acids • Glucose • Ketones	• Glycogen • Creatine phosphate • Muscle proteins	None
Heart muscle	• Fatty acids • Ketones • Glucose	None	None
Adipose	• Glucose • Fatty acids	• TG	• Free fatty acids • Glycerol

f. How do exercising muscles get rid of the $NH4_+$ that builds up from breaking down amino acids into pyruvate and acetyl CoA? (two ways)
   (1) Convert pyruvate → alanine. Then export alanine to the liver.
   (2) Convert glutamate → glutamine. Then export glutamine to the liver.
4. **The four most common catabolic states and the main (but not only) hormones released in each** (Table 8.7)
5. **The most common catabolic hormones and their effects** (Table 8.8)

TABLE 8.7 Four Most Common Catabolic States and the Hormones Released in Each

Catabolic State	Main Hormone(s) Released in Response
Exercise	Epinephrine and norepinephrine from adrenal medulla
Stress	• Epinephrine and norepinephrine from adrenal medulla • Cortisol from adrenal cortex
Fasting	Glucagon from pancreas alpha cells
Cold/exposure	Thyroid hormones from thyroid gland

TABLE 8.8 Regulation and Effects of Common Catabolic Hormones

Catabolic Hormone	Main Factors that ↑ Secretion	Metabolic Effects	Main Inhibitors of Secretion
Glucagon from pancreas alpha cells	**DIRECT** • ↓ Blood [glucose] • ↑ Blood [amino acids] (especially arginine) • CCK • Epinephrine, norepinephrine • Ach **INDIRECT** • Stress (stimulates release of epinephrine and norepi) • Exercise (stimulates release of epinephrine and norepi) • Fasting (lowers blood glucose)	This is a brief summary; more details will follow • ↑ Glycolysis • ↑ Blood [glucose] • ↑ Blood [fatty acid] • ↑ Ketone production • ↓ Blood [amino acid] How? ↑ Amino acid uptake into the liver as a substrate for gluconeogenesis • ↓ Cholesterol synthesis • Stimulates insulin release	• ↑ Blood [glucose] • ↑ Blood [fatty acids] • ↑ Blood [ketone bodies] • Insulin • Somatostatin

Continued

TABLE 8.8 Regulation and Effects of Common Catabolic Hormones—cont'd

Catabolic Hormone	Main Factors That ↑ Secretion	Metabolic Effects	Main Inhibitors of Secretion

Cortisol from adrenal cortex	**DIRECT**
	• CRH from hypothalamus and ACTH from anterior pituitary
	INDIRECT
	(Act by ↑ hypothalamic CRH secretion)
	• Stress
	• Circadian rhythm (cortisol levels peak about 8 AM; probably as a response to overnight fasting)
		• ↑ Blood glucose; how?
		  ▪ ↑ Gluconeogenesis
		  ▪ ↓ Glucose uptake into muscle and fat
		  ▪ Enhances effect of glucagon to breakdown glycogen
		• ↑ Fat breakdown; why?
		  ▪ To use glycerol in gluconeogenesis
		• ↑ Blood (amino acids); how?
		  ▪ ↑ Protein breakdown
		  ▪ ↓ Protein synthesis
		• Why?
		  ▪ To use amino acids in gluconeogenesis
		• ↑ Insulin release
		• Inhibits HGH release
			• Negative feedback by cortisol itself in 3 ways:
			(1) On the hypothalamus—inhibits neural input
			(2) On the hypothalamus—inhibits CRH secretion
			(3) On the anterior pituitary—inhibits ACTH secretion

Epinephrine and norepinephrine from adrenal medulla	**DIRECT**
	• Sympathetic stimulation of adrenal medulla
	INDIRECT
	Act by ↑ sympathetic stimulation of adrenal medulla
	• Exercise
	• Stress
	• Fear
	• Pain
	• Extreme cold/exposure
	• Trauma
	• Hypovolemia (e.g., hemorrhage)
	• Low [oxygen]
	• Low blood [glucose]
		• ↑ Blood glucose; how?
		  ▪ ↑ glycogen breakdown
		  ▪ ↑ Gluconeogenesis
		• ↑ Blood fatty acid; how?
		  ▪ ↑ TG breakdown
		  ▪ ↓ Fatty acid synthesis
		• ↑ Uptake of K^+ into muscles
		• ↑ Ketone formation
		• ↑ Heart and respiration rate
		• ↓ Insulin release
		• ↑ Glucagon release
			• Restoration of homeostasis regarding the stimuli that induce secretion (e.g., normal activity rate, normal temp, normal blood volume, normal blood [oxygen], normal blood [glucose])

Thyroid hormones (T3 and T4) from the thyroid gland	**DIRECT** • TRH from hypothalamus, and TSH from anterior pituitary **INDIRECT** (Act by ↑ hypothalamic TRH secretion) • Cold • Ingestion of excess calories	**PRIMARY EFFECTS** • ↑ Heart and respiration rate • ↑ BMR • ↑ Cellular O_2 consumption • ↑ Bone growth and maturation • ↑ Sensitivity to epinephrine and norepinephrine	• Negative feedback from T3 itself at 2 levels: (1) Hypothalamus (inhibits TRH) secretion (2) Anterior pituitary (inhibits TSH secretion)

Continued

TABLE 8.8 Regulation and Effects of Common Catabolic Hormones—cont'd

Catabolic Hormone	Main Factors That ↑ Secretion	Metabolic Effects	Main Inhibitors of Secretion

Thyroid hormones		**SECONDARY EFFECTS** (results of the primary effects)
		• ↑ Heat production
		• ↑ Metabolism of carbs, fats, and proteins to meet the ↑ energy demands of the ↑ BMR
		  ▪ ↑ Glucose absorption from GI tract
		  ▪ ↑ Glycogen breakdown
		  ▪ ↑ Flux through glycolysis
		  ▪ ↑ Gluconeogenesis
		  ▪ ↑ Fat breakdown
		  ▪ ↑ Protein degradation
			• Dopamine (inhibits TSH secretion)
			• Somatostatin (inhibits TSH secretion)
			• Cortisol (inhibits TSH secretion)

C. ANABOLISM: SYNTHESIZING IMPORTANT MATERIALS
  1. **Anabolic processes categorized by substrate and location** (Table 8.9)
  2. **The three most common anabolic states and main hormones that are released in response** (Table 8.10)
  3. **The three most common anabolic hormones and their effects** (Table 8.11)

D. THREE MAJOR METABOLIC STATES AND THE MAIN REACTIONS OCCURRING IN DIFFERENT TISSUES
  1. **You just ate** (i.e., **absorptive period**). What are the main metabolic processes of which you need to be aware in the...
     a. **Liver?**
        (1) Main energy source = **glucose** (via glycolysis and TCA cycle)
        (2) Synthesis of glycogen, FFAs, TGs, and proteins
        (3) Secretion of VLDL to transport fatty acids and cholesterol to peripheral tissues
     b. **Adipose?**
        (1) Main energy source = **glucose** (via glycolysis and TCA cycle)
        (2) Synthesis of FFA and TG
     c. **CNS?**
        • Main energy source = **glucose** (via glycolysis and TCA cycle)
     d. **Skeletal muscle?**
        (1) Main energy source = **glucose** (via glycolysis and TCA cycle)
        (2) Synthesis of glycogen and proteins
     e. **Heart muscle?**
        • Main energy source = **FFAs** (via beta oxidation and TCA cycle)

TABLE 8.9 Fuel Substrates and the Anabolic Processes They Participate In

Substrate Involved	Anabolic Processes	Location	Description
Carbohydrates	Conversion of monosaccharides to disaccharides	Cytoplasm	Glucose, fructose, or galactose → lactose, maltose, or sucrose
	Glycogen synthesis	Cytoplasm of liver and skeletal muscle	G-6-P → glycogen
	Gluconeogenesis	Cytoplasm of liver and kidney	• Lactate • Pyruvate • OAA • Glycerol phosphate • Glucogenic amino acids • Propionyl CoA → glucose
Fats	De novo synthesis of FFA from acetyl CoA	Cytoplasm of liver, adipose, lactating mammary, and kidney cells	Acetyl CoA → free fatty acids
	TG synthesis	Cytoplasm of liver, adipose, and brush border cells	FFAs + glycerol phosphate → TG
Proteins	Protein synthesis (aka translation)	Cytoplasm of all cells; especially liver and skeletal muscle	Amino acids → proteins
Nucleotides	Nucleotide synthesis	Cytoplasm of all cells	N bases → nucleosides → nucleotides

TABLE 8.10 Most Common Anabolic States and the Hormones Released in Each

Anabolic State	Main Hormone(s) Released in Response
Postabsorptive state	Insulin from pancreas beta cells
Growth spurt	• HGH from the anterior pituitary • Androgens from gonads and adrenal glands
Deep sleep	HGH from the anterior pituitary

TABLE 8.11 Regulation and Effects of Common Anabolic Hormones

Anabolic Hormone	Main Factors that Stimulate Secretion	Metabolic Effects	Main Inhibitors of Secretion
Insulin from pancreas beta cells	• ↑Blood [glucose] • ↑Blood [fatty acid] • ↑Blood [amino acid] • ↑GI hormones (e.g., secretin, GIP, gastrin, CCK) • ↑ACh (i.e., parasympathetic stimulation) • ↑HGH • ↑Cortisol • ↑Glucagon (this is counterintuitive) • ↑Sulfonylurea drugs (e.g., tolbutamide, glyburide)	This is a brief summary. More details about insulin's actions later. • ↓Blood glucose • ↑Glucose, fat, and amino acid storage • ↑Cell growth • ↑Cholesterol synthesis • Inhibits glucagon release	• ↓Blood [glucose] • ↓Blood [fatty acid] • Somatostatin • Epinephrine, norepinephrine

Human growth hormone (HGH) from the anterior pituitary	DIRECT
	• GHRH from hypothalamus
	INDIRECT (these act by ↑GHRH secretion)
	• ↓blood [glucose]
	• ↓blood [fatty acid]
	• ↑blood [amino acids]
	• Fasting
	• Exercise
	• Stage 4 (deep sleep)
	• Growth spurt
	• ↑T3 (a thyroid hormone)
		• ↑Blood glucose; how?
		  ■ ↓Gluconeogenesis
		  ■ ↓Glucose uptake into muscle and adipose
		• ↑Fat breakdown
		• ↑Amino acid uptake into muscles
		• ↑Protein synthesis
		• ↑Linear growth
		• ↑Organ size
		• Stimulates insulin release
			DIRECT
			• Release of somatostatin from the hypothalamus
			• Negative feedback from HGH itself at the hypothalamus (inhibits GHRH release)
			• Negative feedback from IGF-1 at the hypothalamus (inhibits GHRH release) and the anterior pituitary (inhibits HGH release)
			INDIRECT Act by ↑Somatostatin and/or ↓GHRH release from hypothalamus
			• ↑Blood [glucose]
			• ↑Blood [fatty acid]
			• Cortisol
			• Pregnancy

Androgens from the adrenal glands and gonads	DIRECT • GnRH from hypothalamus and LH from anterior pituitary	• ↑Protein and muscle synthesis • ↑Libido	• Negative feedback from testosterone itself at the hypothalamus (inhibits GnRH release) and the anterior pituitary (inhibits LH release)

2. You **have not eaten in hours** (i.e., **postabsorptive period**). What are the main metabolic processes of which you need to be aware in the...
   a. **Liver?**
      (1) Main energy source = **FFAs** (via beta oxidation and TCA cycle)
      (2) Breakdown of glycogen—release of glucose into the blood
      (3) Some gluconeogenesis
   b. **Adipose?**
      (1) Main energy source = **FFAs** (via beta oxidation and TCA cycle)
      (2) Breakdown of TGs—release of FFAs and glycerol into the blood
   c. **CNS?**
      - Main energy source = **glucose** (via glycolysis and TCA cycle)
   d. **Skeletal muscle?**
      (1) Main energy source = **FFAs** (via beta oxidation and TCA cycle)
      (2) Breakdown of glycogen—intracellular usage of resulting glucose
      (3) Some breakdown of muscle proteins to provide amino acids for gluconeogenesis. Also, some deamination of amino acids in order to feed carbon skeletons into the Krebs cycle
   e. **Heart muscle?**
      - Main energy source = **FFAs** (via beta oxidation and TCA cycle)
3. You **haven't eaten in days** (i.e., **starvation**). What are the main metabolic processes of which you need to be aware in the...
   a. **Liver?**
      (1) Main energy source = **FFAs** (via beta oxidation and TCA cycle)
      (2) Synthesis of ketones and release into the bloodstream
      (3) ↑ Gluconeogenesis
      (4) Breakdown of proteins into amino acids, and deamination of these amino acids to use the carbon skeletons: (1) in Krebs, (2) in ketone formation, and (3) in gluconeogenesis
   b. **Adipose?**
      (1) Main energy source = **FFAs** (via beta oxidation and TCA cycle)
      (2) Breakdown of TGs and release of FFAs and glycerol into the blood
   c. **CNS?**
      - Main energy source = **glucose** (via glycolysis and TCA cycle). However, the longer the period of starvation, the greater the usage of ketone bodies.
   d. **Skeletal muscle?**
      (1) Main energy source = **FFAs and ketones** (via beta oxidation and TCA cycle)
      (2) ↑ Breakdown of muscle proteins to provide amino acids for gluconeogenesis, also deamination of certain amino acids in order to feed carbon skeletons into the Krebs cycle
      (3) **Clinical correlation → After how many days of starvation does death occur?** → About 60. **What is the cause of death?** → Breakdown of essential proteins of the heart and brain for energy
   e. **Heart muscle?**
      - Main energy source = **ketones** (via TCA cycle)

E. **Citrate: An important metabolic player**
   There are three important metabolic roles of citrate:
   1. **Source of acetyl CoA for de novo FFA synthesis** (via use of the citrate shuttle).
   2. **Stimulates de novo FFA synthesis** by stimulating acetyl CoA carboxylase in the cytoplasm.
   3. **Shuts down glycolysis** by inhibiting PFK.

F. **Pyruvate: A crucial substrate at a metabolic crossroads**
   1. **The three fates of pyruvate in a <u>nonhepatic</u> cell:**
      a. Under **low O_2 conditions—converts to lactate** in order to regenerate NAD^+. (Enzyme that converts pyruvate to lactate? → Lactate dehydrogenase)
         (1) Lactate can then be sent to the liver for gluconeogenesis.
      b. Under **high O_2 conditions—enters mitochondrial matrix and gets decarboxylated** for entry into the Krebs cycle. (Enzyme that decarboxylates pyruvate? → Pyruvate dehydrogenase)
      c. **Converts to alanine** as a means of getting rid of intracellular NH3. (Enzyme responsible for converting pyruvate to alanine? → Aminotransferase)
         (1) Alanine is then exported and sent to the liver, where it either enters the urea cycle or is used for gluconeogenesis.
   2. **The four fates of pyruvate in a <u>liver</u> cell** (The first two are the same as in a nonhepatic cell; the last two are unique.)
      a. Under **low O_2 conditions—converts to lactate** to regenerate NAD^+. (Enzyme that converts pyruvate to lactate? → Lactate dehydrogenase)
      b. Under **high O_2 conditions—enters mitochondrial matrix and gets decarboxylated** to yield acetyl CoA for entry into the Krebs cycle. (Enzyme that decarboxylates pyruvate? → Pyruvate dehydrogenase)
      c. **Converts to OAA to stimulate flux through the Krebs cycle** (liver and kidney only). (Enzyme that converts pyruvate to OAA? → Pyruvate decarboxylase)
      d. **Converts to OAA to enter gluconeogenesis** (liver and kidney only)

G. **Lactate**
   1. **Where does lactate originate?** → From the last step of anaerobic glycolysis—conversion of pyruvate to lactate by lactate dehydrogenase
   2. **Why is the conversion of pyruvate to lactate crucial for the survival of cells under low O_2 conditions?** → B/c this reaction regenerates the NAD^+ that glycolysis MUST HAVE to continue.
   3. **What are the three fates of lactate in an exercising muscle?**
      a. **Builds up** within the cell
      b. Is exported to the liver and kidneys as a **gluconeogenic precursor** (this is the first part of the "Cori cycle")
      c. Is sent to the **liver and heart muscle** as a **supplementary energy source** (gets converted to pyruvate and enters TCA cycle)

OVERVIEW OF METABOLISM WITH EMPHASIS ON HORMONAL CONTROL

H. Acetyl CoA: Another crucial substrate at a metabolic crossroads

KEY IDEA: Acetyl CoA is a key metabolic intermediate. You must understand its metabolic fates and its effects on certain pathways.

1. Repeat after me: "acetyl CoA cannot be used in gluconeogenesis. Therefore fat cannot be converted into glucose." Good. Now say it again three times. Trust me; this is important.
2. **The three fates of acetyl CoA:**
   a. **Enters Krebs cycle** to generate energy (all cells)
   b. Substrate for **de novo fatty acid synthesis** (only in liver, adipose, lactating mammary, and kidney cells)
   c. Substrate for **ketone body synthesis** (only in liver cells)
3. **What happens when acetyl CoA levels are <u>high</u>?**
   a. ↑ Flux through pathways that use acetyl CoA as a reactant
   b. ↓ Flux through pathways that produce acetyl CoA as a product
   • Note: More details on this are found in Chapter 10: Carbohydrate Metabolism.
4. **What happens when acetyl CoA levels are <u>low</u>?** → The reverse of the above reactions
   a. ↓ Flux through pathways that use acetyl CoA as a reactant
   b. ↑ Flux through pathways that produce acetyl CoA as a product

I. Glucose metabolism: Up close and personal

1. **Quick facts about glucose**
   a. Glucose requirement of the <u>body</u> per day?   160 g
   b. Glucose requirement of the <u>brain</u> per day?   120 g (75% of total body requirement!)
   c. Amount of glucose <u>stored as glycogen?</u>   190 g (i.e., about one day's supply)
2. **What are the only three sources of glucose?**
   a. Diet
   b. Glycogen in the liver and muscles
   c. Gluconeogenesis
3. **What are the two ways that glucose can enter a cell?**
   a. **Facilitated diffusion through glucose transporters.**
      (1) This is the most common method.
      (2) Entry of glucose into the cell is driven by diffusion down its concentration gradient.
      (3) **Problem:** Certain cells have a higher concentration of glucose **inside** than outside. Therefore they cannot rely on the concentration gradient to power glucose uptake. **Solution:** Method #2—Cotransport with Na.
   b. **Cotransported with Na**
      (1) Glucose flows into the cell <u>against</u> its concentration gradient by being coupled with the movement of Na <u>down</u> its concentration gradient.

(2) This occurs in three places:
   (a) Brush border cells of the small intestine (dietary absorption)
   (b) Proximal convoluted tubule cells of renal nephrons (renal reabsorption)
   (c) Epithelial cells of the choroid plexus
4. What <u>stimulates</u> glucose uptake into skeletal muscle and fat cells?
   → Insulin
   a. **How?** → It increases the number and activity of Glut 4 transport proteins in skeletal muscle and in adipose.

KEY IDEA: Glucose uptake into the following three tissues is NOT affected by insulin levels: (1) Brain, (2) liver, and (3) RBCs.

5. What three things <u>inhibit</u> glucose uptake in skeletal muscle and fat cells?
   a. Cortisol
   b. HGH
   c. High blood (FFA)
6. What are the two fates of glucose once it enters a cell?
   a. Enters glycolysis **(or)**
   b. Is stored as glycogen

KEY IDEA: There is only one hormone that has the effect of ↓ blood glucose. Know it. → **Insulin**. How does insulin ↓ blood glucose?
1. By ↑ **glucose uptake** into muscle and fat cells (by recruiting glut 4 receptors)
2. By ↓ **glycogen breakdown** and ↑ **glycogen synthesis** in liver and muscle
3. By ↓ **gluconeogenesis** in liver and kidney

KEY IDEA: These four hormones all have the effect of ↑ blood glucose. Memorize them (Table 8.12).

TABLE 8.12 Four Hormones that Increase Blood Glucose

Hormone	How Does It ↑ Blood Glucose?
Glucagon from pancreas alpha cells	(1) ↑ Glycogen breakdown and ↓ glycogen synthesis in liver (but NOT in muscle) (2) ↑ Gluconeogenesis in liver and kidney
Cortisol from adrenal cortex	(1) Enhances the effect of glucagon in degrading glycogen (2) ↑ Gluconeogenesis (3) ↓ Glucose uptake into muscle and fat cells
Epinephrine and norepinephrine from adrenal medulla	(1) ↑ Glycogen breakdown (2) ↑ Glycogen breakdown luconeogenesis
HGH from anterior pituitary	(1) ↑ Gluconeogenesis (2) ↑ Glucose uptake into muscle and fat cells

7. **Flux through glycolysis and gluconeogenesis is influenced by the carbohydrate content of the diet and is mediated by insulin and glucagon.**
   a. High (carbohydrates) = ↑ blood glucose = ↑ insulin = ↑ flux through glycolysis and ↓ flux through gluconeogenesis
   b. Low (carbohydrates) = ↓ blood glucose = ↑ glucagon = ↓ flux through glycolysis and ↑ flux through gluconeogenesis

J. INSULIN: THE BODY'S MEANS OF STORING FUEL SUBSTRATES IN RESPONSE TO THE "WELL-FED" (I.E., ABSORPTIVE) STATE
   1. **How does the body take advantage of the well-fed state?** → Releases insulin
   2. **What serves as a marker for endogenous insulin release?** → C peptide

---

KEY IDEA: Effects of insulin on blood levels of important materials:
1. ↓ **Blood [glucose].** How?
   a. ↑ Glucose uptake into muscle and fat cells
   b. ↑ Glycogen storage, and ↓ glycogen breakdown in liver and muscle
   c. ↓ Gluconeogenesis in liver and kidney
2. ↓ **Blood [amino acids].** How?
   a. ↑ Amino acid uptake into muscle and liver
   b. ↑ Protein synthesis in muscle and liver
   c. ↓ Protein breakdown in muscle and liver
3. ↓ **Blood [fatty acid].** How?
   a. ↑ Fatty acid uptake into adipose and liver cells
   b. ↑ Triglyceride synthesis in adipose and liver cells
   c. ↓ Triglyceride breakdown in adipose and liver cells
4. ↓ **Blood [K$^+$, magnesium, and phosphate]**
   a. **How?** ↑ Uptake of K$^+$, magnesium, and phosphate into muscle and fat cells.
   b. **Why?** Growing cells have increased requirements of these minerals.
   c. **Note:** This effect of insulin can potentially create hypokalemia, hypomagnesemia, and hypophosphatemia.
5. ↓ **Blood [ketone]**
   a. **How?** ↓ TG breakdown in the liver
   b. **What is the connection between** ↓ **TG breakdown in the liver and** ↓ **blood [ketone]?** → The following pathway: ↑ insulin → ↓ TG breakdown → ↓ (FFA) → ↓ beta oxidation → ↓ acetyl CoA → ↓ ketone formation
   c. **Note:** The main stimulus of ketone body formation is acetyl CoA levels.
6. ↓ **Blood [glucagon]. How?** → Inhibits glucagon secretion from pancreas alpha cells

---

   3. **Insulin: Summary of its physiologic effects and mechanisms** (Table 8.13)
   4. **Effects of insulin categorized by target tissue** (Table 8.14)
   5. **Insulin: Regulation of secretion** (Table 8.15)

Table 8.13 Insulin: Effects and Mechanisms

Summary of its Physiologic Effects	Target Tissue	Biochemical Mechanism
Turn glucose into energy (since you have got plenty of it around)	Muscle and adipose All cells	↑ Uptake of glucose ↑ Glycolysis
Do not waste energy creating glucose or ketones (since you have plenty of glucose around)	Liver and kidney Liver	↓ Gluconeogenesis ↓ Ketone formation (as a result of ↓ TG breakdown)
Store glucose	Liver and muscle	↑ Glycogen synthesis, ↓ glycogen breakdown
Synthesize proteins	Muscle and liver	↑ Amino acid uptake, ↑ protein synthesis, ↓ protein breakdown, ↑ transcription, ↑ K+, magnesium, and phosphate uptake Why? B/c growing cells have an ↑ requirement for these minerals
Store fat	Adipose and liver Liver	↑ FFA uptake, ↑ de novo FFA synthesis, ↑ triglyceride synthesis, ↓ triglyceride breakdown ↓ Beta oxidation of FFA
Synthesize cholesterol	All cells	↑ Cholesterol synthesis
Stop your enemy hormone from working against you!	Pancreas alpha cells	Inhibits glucagon secretion

Table 8.14 Effects of Insulin Categorized by Target Tissues

Target Tissues	Effects
All cells	↑ Glycolysis, ↑ cholesterol synthesis
Muscle	↑ Glucose uptake, ↑ glycogen synthesis, ↓ glycogen breakdown, ↑ amino acid uptake, ↑ protein synthesis, ↓ protein breakdown, ↑ uptake of K+, magnesium and phosphorus
Kidneys	↓ Gluconeogenesis
Liver	**(Carbs)**—↑ glycogen synthesis, ↓ glycogen breakdown, ↓ gluconeogenesis **(Fats)**—FFA uptake, ↑ de novo FFA synthesis, ↓ beta oxidation of FFA, ↑ TG synthesis, ↓ TG breakdown **(Ketones)**—↓ ketone formation **(Proteins)**—↑ amino acid uptake, ↑ protein synthesis, ↓ protein breakdown, ↑ uptake of K+, magnesium and phosphorus
Adipose	↑ Glucose uptake, ↑ FFA uptake, ↑ de novo FFA synthesis, ↑ TG synthesis, ↓ TG breakdown
Pancreas alpha cells	Inhibit glucagon secretion

Table 8.15 Regulation of Insulin Secretion

Where is insulin produced?	Beta cells of pancreas	
Summary of insulin's physiologic effects	(1) ↓ Blood glucose (2) ↑ Fuel substrate storage (3) ↑ Cell growth (4) Inhibit glucagon release	
What stimulates insulin release?	**DIRECT stimulators of insulin release** • ↑ Blood [glucose] (e.g., after a meal)	**Why does this make sense? What is your body thinking?** • Since you have lots of glucose around, you should take advantage by using it for energy and storing it.
	• ↑ Blood [amino acids] (e.g., after a meal)	• Since you have lots of amino acids around, you should take advantage by using them to synthesize proteins.
	• ↑ Blood [fatty acids]	• Since you have fatty acids around, you should take advantage by storing them and using them to synthesize triglycerides.
	• GI tract hormones (e.g., secretin, GIP, gastrin, CCK)	GI hormones = food in GI tract = impending influx of nutrients that will need to be stored.
	• Acetycholine	• A parasympathetic neurotransmitter ("rest and digest")
	• Sulfonylurea drugs (e.g., tolbutamide, glyburide)	• These cause insulin release and are used to treat diabetes
	• Glucagon	• Hmmmm. Why should glucagon stimulate insulin release? This seems counterintuitive since these hormones have basically opposite effects. However, it is actually a form of negative feedback for glucagon, since insulin inhibits glucagon release. (↑ glucagon → ↑ insulin → ↓ glucagon).
	INDIRECT stimulators of insulin release • Growth hormone (HGH)	**Why does this make sense? What is your body thinking?** This is an indirect effect. HGH → ↑ blood [glucose] → insulin release.
	• Cortisol	This is an indirect effect. Cortisol → ↑ blood [glucose] → insulin release.
What inhibits insulin release?	**DIRECT inhibitors of insulin release** • ↑ Low blood glucose levels	**Why does this make sense? What is your body thinking?** • A further ↓ in blood glucose by insulin would harm the brain (since its primary energy source is blood glucose).
	• ↓ Low blood amino acid levels	
	• ↓ Epinephrine, norepinephrine	Is part of the sympathetic response ("fight or flight"). Under such conditions, breakdown of fuel molecules, not their storage, is needed.
	• ↓ Somatostatin	

6. **How does insulin…?** (Table 8.16)
   - **Note:** Not all of the mechanisms of insulin's actions are explained. The ones that are omitted are either not clearly understood or not as important.

> **KEY IDEA:** What is the second messenger system by which insulin exerts all its effects? → Tyrosine kinase receptors.
> - Remember: Insulin = tyrosine kinase receptor = ↓ (cAMP) = DEphosphorylates enzymes.

TABLE 8.16 How Does Insulin…?

↑ Glucose uptake?	• Promotes insertion of glut four transporters into cell membranes

↑ Glycolysis?	• Increases synthesis of 3 glycolytic enzymes in the liver:
	  ▪ Glucokinase
	  ▪ PFK-1
	  ▪ Pyruvate kinase
	• Stimulates PFK-1 (via ↑ production of F26BP)
	• Reduces inhibition of pyruvate kinase in the liver (via ↓ production of protein kinase A)

↑ Glycogen synthesis?	• Activates glycogen synthase (via dephosphorylation)
↓ Glycogen breakdown?	• Deactivates glycogen phosphorylase (via phosphorylation)
↓ Gluconeogenesis?	• Decreases synthesis of 4 gluconeogenic enzymes (in the liver and kidneys) • Pyruvate carboxylase • PEPCK • Fructose-1,6-bisphosphatase • Glucose-6-phosphatase (in liver only) • Inhibits fructose-1,6-bisphosphatase (via ↑ production of F26BP)
↑ Protein synthesis?	• Stimulates transcription and translation
↑ Free fatty acid uptake?	• ↑ Activity of lipoprotein lipase on cell surfaces
↑ Cholesterol synthesis?	• Stimulates HMG CoA reductase
↓ TG breakdown into FFA and glycerol?	• Inhibits hormone sensitive lipase
↑ De novo FFA synthesis from acetyl CoA?	• ↑ Levels of fatty acid synthase and acetyl CoA carboxylase
↓ Beta oxidation of FFAs in the liver?	• Inhibits transfer of fatty acids into the mitochondria
↓ Ketone formation?	• ↓ Acetyl CoA b/c of ↓ TG breakdown

KEY IDEA: Insulin regulates the expression of its own receptors.
ex) High insulin levels (e.g., obesity) = ↓ # of insulin receptors
ex) Low insulin levels (e.g., starvation) = ↑ # of insulin receptors

K. GLUCAGON: RELEASES FUELS INTO THE BLOOD AND SEQUESTERS AMINO ACIDS IN THE LIVER FOR GLUCONEOGENESIS

KEY IDEA: Effects of glucagon on blood levels of important materials:
1. ↑ **Blood [glucose].** How?
    a. ↑ Glycogen breakdown in the liver
    b. ↓ Glycogen synthesis in the liver
    c. ↑ Gluconeogenesis in liver and kidney
2. ↑ **Blood [fatty acid].** How?
    a. ↑ TG breakdown in liver, slight ↑ in adipose
    b. ↓ TG synthesis in liver and adipose
3. ↑ **Blood [ketone].** How?
    a. ↑ TG breakdown in the liver
    b. **What is the connection between ↑ TG breakdown in the liver and ↑ blood [ketone]?** The following pathway: ↑ glucagon → ↑ TG breakdown → ↑ (FFA) → ↑ beta oxidation → ↑ acetyl CoA → ↑ ketone formation
    c. Remember that the main stimulus of ketone body formation is acetyl CoA levels.
    d. ↓ **Blood [amino acid]**
       • How? ↑ Amino acid uptake into the liver and kidney
       • Why? As a substrate for gluconeogenesis
    e. ↑ **Blood [insulin]**
       • How? Stimulates pancreas beta cells to release insulin
       • Why? As a mechanism of negative feedback. Glucagon stimulates insulin release; which inhibits further glucagon release.

1. **Glucagon—summary of its physiologic effects and mechanisms** (Table 8.17)
2. **Effects of glucagon categorized by target tissue** (Table 8.18)
3. **Glucagon: Summary of its production, stimulation, and inhibition** (Table 8.19)
4. **How does glucagon...?** (Table 8.20)
    • **Note:** Not all of the mechanisms of glucagon's actions are explained. The ones that are omitted are either not clearly understood or not as important.

KEY IDEA: What is the second messenger system by which glucagon exerts all its effects? → **G protein-linked cell receptors.**
• Remember: Glucagon = G protein linked receptor = ↑ (cAMP) = ↑ protein kinase A = phosphorylates enzymes.

Table 8.17 Glucagon: Effects and Mechanisms

Summary of its Physiologic Effects	Target Tissue	Biochemical Mechanism
Stop using glucose for energy (since blood glucose levels are low)	Liver	↓ Glycolysis
Release glucose into the bloodstream (since blood glucose levels are low)	Liver Liver and kidneys	↑ Glycogen breakdown ↓ Glycogen synthesis ↑ Gluconeogenesis ↑ Uptake of amino acids ↑ Deamination of amino acids to yield carbon skeletons for gluconeogenesis
Create an alternative energy source in case there is not enough glucose	Liver	↑ Ketone formation
Encourage the use of fat for energy (instead of glucose)	Liver and adipose	↑ Beta oxidation, ↓ De novo FFA synthesis, ↑ TG breakdown, ↓ TG synthesis
Stop cholesterol synthesis	All cells	Inhibits HMG CoA reductase

Table 8.18 Effects of Glucagon Categorized by Target Tissues

Target Tissues	Effects
All cells	↓ Cholesterol synthesis
Muscle	Nothing! (Note that glucagon does NOT stimulate muscle glycogen breakdown)
Liver cells	**(Carbs)**—↓ glycogen breakdown, ↑ glycogen synthesis, ↓ glycolysis, ↑ gluconeogenesis **(Fats)**—↑ beta oxidation, ↓ de novo FFA synthesis, ↑ TG breakdown, ↓ TG synthesis **(Ketones)**—↑ ketone formation **(Proteins)**—↑ Uptake of amino acids, ↑ deamination of amino acids to yield carbon skeletons for gluconeogenesis
Kidney	↑ Gluconeogenesis, ↑ uptake of amino acids, and ↑ deamination of amino acids to yield carbon skeletons for gluconeogenesis
Adipose	↑ Beta oxidation, ↓ de novo FFA synthesis, ↑ TG breakdown (slight), ↓ TG synthesis
Pancreas beta cells	Stimulates insulin secretion

TABLE 8.19 Regulation of Glucagon Secretion

Where is glucagon produced?	Alpha cells of pancreas	
What stimulates glucagon release?	**DIRECT stimulators of glucagon release** • ↓ Blood [glucose]	**Why does this make sense? What is your body thinking?** Since the major function of glucagon is to ↑ blood glucose, low blood glucose is a logical stimulus for glucagon secretion.
	• ↑ Blood [amino acid] (e.g., after a meal)	Remember that ↑ blood [amino acid] also stimulates insulin release. However, too much insulin could cause hypoglycemia. To avoid provoking insulin-induced hypoglycemia, blood amino acids simultaneously stimulate glucagon release.
	• Epinephrine and norepinephrine	In a time of stress you need ↑ levels of blood glucose to provide energy for the brain and muscles.
	• CCK	One stimulus for CCK release is small peptides in the duodenum. Thus, CCK is another means by which amino acids stimulate glucagon release.
	• Acetylcholine	
What inhibits glucagon release?	**DIRECT Inhibitors of glucagon release** • ↑ Blood [glucose]	**Why does this make sense? What is your body thinking?** Since the major function of glucagon is to ↑ blood glucose, high blood glucose is a inhibitor of glucagon secretion.
	• ↑ Blood [fatty acid]	Since a function of glucagon is to ↑ blood FFA, high blood FFA is a logical inhibitor of glucagon secretion.
	• ↑Blood [ketone]	Since a function of glucagon is to ↑ ketone production, high [ketone] is a logical inhibitor of glucagon secretion.
	• Insulin	Insulin has several effects than oppose those of glucagon. So why simultaneously secrete two hormones that have opposing effects? You should not! Thus insulin acts to inhibit glucagon release.
	• Somatostatin	

Table 8.20 How Does Glucagon...?

↑ Glycogen breakdown and ↓ glycogen synthesis in the liver?	• Activates glycogen phosphorylase • Inhibits glycogen synthase

↓ Glycolysis?	• Decreases synthesis of three glycolytic enzymes in the liver
	    ▪ Glucokinase
	    ▪ PFK-1
	    ▪ Pyruvate kinase
	• Reduces stimulation of PFK-1 (via ↓ production of F26BP)
	• Inhibits pyruvate kinase in the liver (via ↑ production of protein kinase A)

↑ Gluconeogenesis?	• Increases synthesis of three gluconeogenic enzymes (in the liver and kidneys) • PEPCK • Fructose-1,6-bisphosphatase • Glucose-6-phosphatase • Reduces inhibition of F16-bisphosphatase (via ↓ production of F26BP) • Inhibits the glycolytic enzyme pyruvate kinase (which causes a buildup of PEP, which is then shunted into gluconeogenesis)
↑ TG breakdown?	• Stimulates hormone sensitive lipase
↓ TG synthesis?	• Decreases cytosolic [FFA] (by stimulating FFA movement into the mitochondria)
↓ De novo FFA synthesis from acetyl CoA?	• Deactivates acetyl CoA carboxylase

Table 8.21 Insulin and Glucagon Have Opposing Effects on the Blood Levels of Three Substrates

Substrate Present in the Blood	Effect of Insulin	Effect of Glucagon
Blood (glucose)	↓	↑
Blood (FFA)	↓	↑
Blood (ketone)	↓	↑

L. **SUMMARY: INSULIN AND GLUCAGON HAVE OPPOSING EFFECTS ON THE BLOOD LEVELS OF THREE IMPORTANT SUBSTRATES** (Table 8.21)

M. **THE LIVER: IT DOES EVERYTHING BUT IRON YOUR SOCKS**

    Important functions of the liver:
    1. **Carbohydrate and glucose metabolism**
        a. Converts all incoming dietary monosaccharides into glucose
        b. Main site of glycogen synthesis and storage
        c. Only site of glycogen that can release glucose into the blood
    2. **Protein metabolism**
        a. Handles all synthesis of nonessential amino acids
        b. Synthesis of all major plasma proteins
            ex) albumins, globulins, fibrinogens

3. **Fat metabolism**
   a. Enables fat digestion by producing bile
   b. Major site of FFA and TG synthesis (some also in adipose cells)
   c. Creates glycerol-phosphate (which is necessary for TG synthesis)
4. **Production of fuel substrates during starvation**
   a. Only site of ketone formation
   b. Main site of gluconeogenesis (90%) (10% in kidneys)
      - As part of gluconeogenesis, it participates in the Cori cycle and the glucose/alanine cycle.
5. **Ammonia (NH_4^+) metabolism**
   - Clearance of ammonia (which is toxic to CNS) by conversion to urea in the urea cycle
6. **Cholesterol homeostasis**
   a. Major site of cholesterol synthesis and transport to peripheral tissues via LDL
   b. Prepares cholesterol for excretion by conversion into bile
7. **Vitamin and mineral metabolism**
   a. Major storage site of the following vitamins: A, D, B_{12}
   b. Second most important storage site of iron (The #1 site is hemoglobin in RBCs)
8. **Detoxify the blood**
   - p450 enzyme metabolism of drugs and/or poisons
9. **Immune system**
   - Important for immune defense against various antigens
10. **Blood production**
    a. Site of destruction of old RBCs (along with the spleen) and recycling of the iron atoms that get released
    b. Site of heme synthesis (along with the bone marrow)
    c. Acts as a reservoir of blood during trauma (in response to hemorrhage, contracts and releases blood into circulation in an effort to restore blood volume)

N. THE BRAIN: A METABOLICALLY UNIQUE TISSUE

How is the brain metabolically unique? (three main ways)
1. **It has only two energy sources:**
   a. Glucose
   b. Ketone bodies (from the liver)
      - **Why such a limited range of fuel substrates?**
      (1) Has no stored glycogen
      (2) Has no significant stores of TGs
      (3) Amino acids in the blood have trouble crossing the blood brain barrier
      (4) FFA in the blood have trouble crossing the blood brain barrier
      (5) Body is not willing to break down brain proteins for energy
2. **The brain's uptake of glucose is independent of insulin levels.**

3. **The brain uses about 75% of the body's daily glucose requirement!**
   a. Total glucose requirement of 160-kg man per day = 160
   b. Glucose requirement of brain per day = 120 g (75%!)
   c. **Why does the brain use so much energy?** → To fuel the Na+/K+ pumps necessary to maintain neuron membrane potentials.
   d. **Clinical correlation** → Because the brain uses so much energy, it is the first organ adversely affected by anything that impairs cellular energy generation (e.g., inhibition of the electron transport chain resulting from a lack of O_2 or poisoning).

Now that you know all there is to know about metabolism, what does it matter? You know what happens when everything works right. What about when things are messed up? See if you can figure out the following clinical pathologies.

HPI: 38 yo woman presents to the ER after being found in a coma. She had complained of dizziness when she went without eating for prolonged durations. On PE, tachycardia (fast heart beat) is noted, along with cold, clammy and sweaty hands. An MRI (magnetic resonance imaging) shows a **mass in the tail of the pancreas**. An infusion of glucose aids the patient out of her coma. **Elevated immunoreactive C-peptide** is noted. What is her problem?
→ Insulinoma.
- A beta cell pancreatic islet cell tumor is causing too much insulin release. Can be associated with **MEN**s (multiple endocrine neoplasias). Immunoreactive C-peptide will be elevated in tumors, whereas it will be decreased with exogenous insulin administration (a psych or iatrogenic issue). It's true—some people inject themselves with insulin to pass out!

So what happens when glucose levels are too high?

HPI: A 45 yo man is brought to the ER in a comalike state. The patient is a known type II diabetic who was known to be compliant with his medication. Two weeks ago, he was hospitalized for a severe pneumonia. On PE, he is found to be hypotensive, tachycardic, and severely dehydrated. MSE (mental status examination) reveals a patient who is confused but gets a 12 on the GCS (Glasgow coma score) out of 15. His CBC (blood work) is normal except for a slightly elevated leukocyte count. Blood glucose level is **800 mg/dL** (high). The serum **osmolality is calculated to be 400 mOsm/kg** (high). Glycosuria **without ketonuria** is noted on the urinalysis. Head CT (computed tomography) is normal. What is the name of this problem?
→ Hyperosmolar hyperglycemic nonketotic coma. (We basically spelled it out for you above!)

What are associated diseases that relate to high glucose?

> **HPI:** 57 yo man presents with blurry vision, headache, enlarging and coarsening facial skin, fingers, and hand. He has also noted that his voice has been deepening and his head appears to be "getting bigger." PE reveals hypertension, prominent facial features including jaw, chin, and an apparently bulging forehead. Large, coarse hands and feet are noted. He cannot see peripherally and, on confrontational testing confirmed by visual field testing, a bitemporal hemianopsia is diagnosed. A high blood glucose level is noted. Brain imaging shows a classic lesion. Make that diagnosis, champ...!
> → **Pituitary adenoma** → increased HGH → acromegaly (in adults); gigantism (in children)

Pop quiz: How does HGH cause high blood glucose? (Can you remember?)
1. It ↑ gluconeogenesis.
2. It ↓ glucose uptake into muscle and fat cells.

Hyperglycemia can present in a variety of clinical situations. Be aware of the many possible causes in making your differential diagnosis.

Chapter 9
TCA Cycle and Electron Transport Chain

A. **The TCA cycle: Generating NADH and $FADH_2$ to send to the electron transport chain (ETC)**
   1. **Location:** The TCA cycle occurs in the mitochondrial matrix of all cells except RBCs
   2. **Functions of the TCA cycle in the body:** Why is the TCA cycle necessary for the body? (six main reasons)
      a. It is **the major mechanism for cellular energy generation** (via creation of GTP, NADH, $FADH_2$).
      b. It provides **a way for carbohydrates, fats, and amino acids to be converted into energy.**
         (1) Carbohydrates enter glycolysis and produce pyruvate, which is converted to Acetyl-CoA, which enters the TCA cycle.
         (2) FFAs enter beta oxidation and produce acetyl CoA, which enters the TCA cycle.
         (3) Amino acids are deaminated to yield carbon skeletons, which enter the TCA cycle.
      c. It provides **a way for certain amino acids to participate in gluconeogenesis.** Certain amino acids yield propionyl CoA, which can be converted to succinyl CoA, which can enter TCA and be converted to OAA, which can enter gluconeogenesis.
      d. It provides **building blocks for amino acid synthesis.**
         What are the TCA cycle intermediates that are removed and used as carbon skeletons in amino acid synthesis?
         (1) Pyruvate (gets transaminated and forms alanine)
         (2) OAA (gets transaminated and forms aspartate)
         (3) Alpha-KG (gets transaminated and forms glutamate)
      e. It provides **citrate, which is used in de novo synthesis of fatty acids.**
      f. It provides **building blocks for heme synthesis.**

Key idea: The TCA cycle begins with acetyl CoA and generates NADH, $FADH_2$, and GTP. It is crucial for the body because it produces more energy than any other metabolic pathway.

3. **How is the TCA cycle connected to the electron transport chain?** → The TCA cycle generates NADH and $FADH_2$, which are then sent to the electron transport chain to generate ATP.
4. **Metabolic context:** The TCA cycle is always occurring. It is a constant and indispensable source of energy. If it were to stop functioning, you would die within minutes.
5. **Overview of the TCA cycle:** Pyruvate or acetyl CoA + OAA + NAD + FADH + GDP →→ OAA + CO_2 + NADH + $FADH_2$ + GTP
   a. **The two starting reactants and where they come from:**
      (1) **Pyruvate** comes from three sources:
         (a) Glycolysis
         (b) Deamination of certain amino acids
         (c) Conversion from lactate
            - **How does pyruvate enter the mitochondria?** → By facilitated diffusion.
      (2) **Acetyl CoA** comes from four sources:
         (a) Conversion from pyruvate
         (b) Beta oxidation of FFAs
         (c) Catabolism of ketone bodies
         (d) Deamination of the amino acid isoleucine
      (3) **Note:** Some molecules join the TCA cycle somewhere in the middle. For example, certain amino acids yield alpha-KG. Certain others yield succinyl CoA.
   b. **The four end-products and where they go:**
      (1) CO_2 (diffuses into the bloodstream for transport and exhalation in the lungs)
      (2) **NADH** (goes to the ETC—electron transport chain)
      (3) **$FADH_2$** (goes to the ETC)
      (4) **GTP** (used by the cell in various ways)
      (5) **Note:** Many of the TCA cycle intermediates can leave the cycle and be used in different ways.
         ex) OAA can enter gluconeogenesis (in liver and kidney only).
         ex) OAA can be used as a carbon skeleton to form aspartate.
         ex) Alpha-KG can be used as a carbon skeleton to form glutamate.
6. **Although the TCA is a cycle, think of it as beginning and ending with OAA (oxaloacetate).**
7. **Keeping track of important molecules produced**
   a. **How many ATP are generated in the step pyruvate → acetyl CoA?** → 3
   b. **How many ATP are generated if acetyl CoA flows completely through the TCA cycle?** → 12
   c. **So, how many ATP are generated after one pyruvate flows completely through the TCA cycle?** → 15 (Pyruvate → acetyl CoA = 3 ATP; acetyl CoA → OAA = 12 ATP)
   d. **One acetyl CoA results in how many turns of the Krebs cycle?** → 1

e. **One acetyl CoA generates the following:**
   (1) 12 ATP
   (2) 3 NADH
   (3) 1 $FADH_2$
   (4) 2 CO_2
   (5) 1 GTP
f. **Note:** Double everything just mentioned per glucose molecule (because glucose generates two acetyl CoA (Figure 9.1).

8. **Key enzymes and their reactions**
   a. **Rate limiting step of TCA?** → Isocitrate dehydrogenase
   b. **Decarboxylation of pyruvate to yield acetyl CoA?** → Pyruvate dehydrogenase
      - **Five coenzymes required by pyruvate dehydrogenase to do this reaction?**
        (1) TPP (a.k.a., thiamin, a.k.a., vitamin B_1)
        (2) FAD (a.k.a., vitamin B_2)
        (3) NAD (a.k.a., niacin, a.k.a., vitamin B_3)
        (4) CoA (a.k.a., pantothenate, a.k.a., vitamin B_5)
        (5) Lipoic acid
   c. **Nutritional correlation** → Notice that four of the five coenzymes required for this reaction are associated with B vitamins! This is one reason that deficiencies of B vitamins cause inhibition of cellular energy production. Observe the following pathway: Lack of B vitamins → ↓ function of pyruvate dehydrogenase → ↓ flux through TCA → ↓ energy production

Do we care about PD? Here's what happens when it's not around:

> **HPI:** 3 yo child presents with severe lethargy, poor feeding, tachypnea, markedly exacerbated during times of illness, stress, or high carbohydrate intake. His symptoms began at birth, when he was noted to have decreased muscle tone, and have progressively worsened. Blood work reveals a lactic acidosis. Imaging shows gray matter degeneration, mistaken by a neurologist for Leigh's disease. A full biochemical workup reveals elevated pyruvate, lactate, and alanine. Enzymatic assays reveal the diagnosis of:
> → PDCD, pyruvate dehydrogenase deficiency
> - Although rare, it is one of the most common of neurodegenerative mitochondrial enzymatic disorders.

   d. **The decarboxylation of alpha-KG to yield succinyl CoA?** → Alpha-KG dehydrogenase
      - **The five coenzymes required by alpha-KG dehydrogenase to do this reaction?** → The same five as for pyruvate dehydrogenase:
        (1) TPP (a.k.a., thiamin, a.k.a., vitamin B_1)
        (2) FAD (a.k.a., vitamin B_2)
        (3) NAD (a.k.a., niacin, a.k.a., vitamin B_3)
        (4) CoA (a.k.a., pantothenate, a.k.a., vitamin B_5)
        (5) Lipoic acid

TCA Cycle and Electron Transport Chain   99

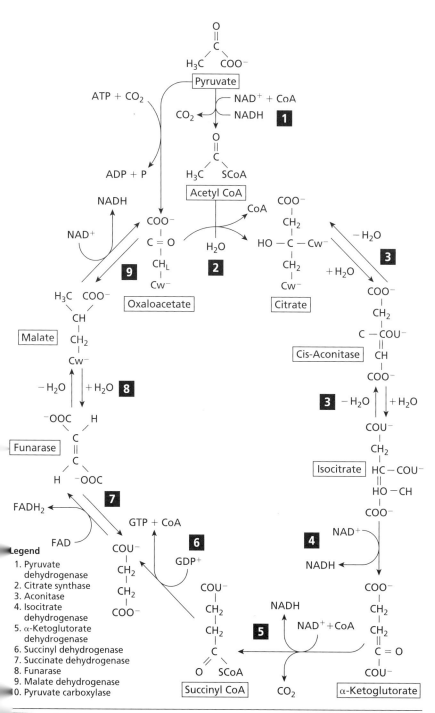

Figure 9.1 TCA cycle.

e. **Nutritional correlation** → Notice again that four of the five coenzymes required for this reaction are B vitamins! This is <u>another reason</u> that deficiencies of B vitamins cause inhibition of cellular energy production. Observe the following pathway: Lack of B vitamins → ↓ function of <u>alpha-KG dehydrogenase</u> → ↓ flux through TCA → ↓ energy production. Notice that the pyruvate dehydrogenase complex is extremely similar to the alpha-KG dehydrogenase complex:
   (1) The same five coenzymes
   (2) The same action (decarboxylation)
   (3) Similar substrates (alpha-KG and pyruvate)
f. **Three decarboxylation steps?**
   (1) Pyruvate dehydrogenase
   (2) Isocitrate dehydrogenase
   (3) Alpha-KG dehydrogenase
g. **Four steps that generate NADH?**
   (1) Pyruvate dehydrogenase
   (2) Isocitrate dehydrogenase
   (3) Alpha-KG dehydrogenase
   (4) Malate dehydrogenase
h. **The one substrate level phosphorylation (creates GTP)?** → Succinate thiokinase
i. **The only step that generates $FADH_2$?** → Succinate dehydrogenase. **What inhibits this reaction?** Malonate (a structural analog of succinate)
j. **The step that liver or kidney cells use to create OAA in order to keep the TCA cycle running at maximum capacity?**
   (1) Pyruvate carboxylase (pyruvate + CO_2 + ATP → OAA)
   (2) Occurs in liver and kidney cells only!!!
k. **Besides helping the TCA cycle run at full capacity, why else is the conversion of pyruvate to OAA helpful?** → It allows pyruvate to enter gluconeogenesis.
l. **What three substances are needed by pyruvate carboxylase to catalyze the conversion of pyruvate → OAA?**
   (1) CO_2
   (2) Biotin
   (3) ATP
9. **Stimulation and inhibition of the TCA cycle**

KEY IDEA: **What are the four ways that the TCA cycle is regulated?**
1. By the **availability of OAA and acetyl CoA**
   - If levels of either OAA or acetyl CoA are low, then flux through TCA slows. When levels of either are excessive, flux through TCA increases.
2. By the **availability of NAD^+ and FAD**
   - If levels of NAD or FAD are low, then flux through TCA decreases. When levels are high, flux through TCA increases.
3. By the **availability of TCA cycle intermediates**
   ex) If alpha KG or OAA levels drop because they are being used as carbon skeletons to synthesize amino acids, then flux through the TCA cycle slows.

KEY IDEA:—CONT'D
4. **By allosteric regulation of specific enzymes.** (For a full discussion, see the chart that follows)
   ex) ATP inhibits pyruvate dehydrogenase
   ex) NADH inhibits isocitrate dehydrogenase

KEY IDEA: What is the <u>single-most powerful inhibitor</u> of the TCA cycle?
→ Acetyl CoA. How? → By inhibiting pyruvate dehydrogenase.

---

10. There are five enzymes in the TCA cycle which are allosterically regulated (Table 9.1).

TABLE 9.1

Reaction	Stimulated by	Inhibited by
Pyruvate dehydrogenase (decarboxylates pyruvate into acetyl CoA in mitochondrial matrix)	• ADP (causes dephosphorylation, thus stimulation, of pyruvate dehydrogenase)	**Direct inhibitors** • Acetyl CoA • ATP (causes phosphorylation, thus inhibition, of pyruvate dehydrogenase) • NADH **Indirect inhibitors** (via production of acetyl CoA) • Fatty acids (produce acetyl CoA via beta oxidation) • Ketones (get catabolized to yield acetyl CoA)
Citrate synthase		• ATP • NADH • Succinyl CoA • Acyl CoA
Isocitrate dehydrogenase	• ADP	• ATP • NADH
Alpha-KG dehydrogenase		• ATP • GTP • NADH • Succinyl CoA
Succinate dehydrogenase		• Malonate (a structural analog of succinate)

B. THE ELECTRON TRANSPORT CHAIN: TURNING NADH AND FADH$_2$ INTO ATP

1. **Generating ATP using the electron transport chain: A two-step approach called "oxidative phosphorylation."** Basically, this is a two-step strategy:
   a. **Step 1: Create a buildup up H$^+$ in the intermembrane space. How?**
      (1) NADH or FADH$_2$ drop off H atoms.

(2) <u>The electrons are removed from the atoms!!!</u> (The electrons flow down the ETC, and the H^+ ions are released into the matrix.)
(3) As electrons flow down the ETC, the free energy release is used by each complex to pump H^+ <u>from</u> the matrix <u>into</u> the intermembrane space. The result is a buildup of H^+ in the intermembrane space. (This is called the "proton gradient.")

<u>**Key idea:** **The mitochondrial inner membrane is impermeable to H^+.** This allows for the creation and maintenance of a high $[H^+]$ in the intermembrane space. Anything that makes the inner membrane permeable to H^+ (e.g., 2,4-DNP) destroys the proton gradient.</u>

b. **Step 2: Exploit the desire of H^+ to flee its area of high concentration by allowing it <u>only one means of escape</u>—being shunted through ATP synthetase.**
   (1) Under normal conditions, the <u>only way</u> H^+ can flee from its high intermembrane concentration is by flowing through ATP synthetase. As H^+ flows down its electrochemical gradient through ATP synthetase, the free energy release is used to phosphorylate ADP.
   (2) **Bottom line:** → The body exploits the flow of H^+ down its concentration gradient by using it to drive the synthesis of ATP (Figure 9.2).

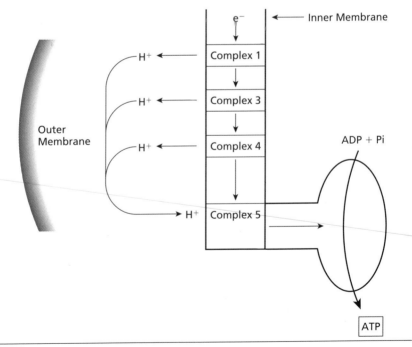

Figure 9.2 Oxidative phosphorylation.

> **KEY IDEA: How many ATP do NADH and FADH$_2$ generate at the ETC?**
> 1. 1 NADH = 3 ATP
> 2. 1 FADH$_2$ = 2 ATP
> 3. **Why the difference?** → B/c NADH and FADH$_2$ drop off their electrons at different points along the ETC. NADH drops off at complex 1; FADH$_2$ at complex 2.

2. **Key principles that apply to all the complexes in the ETC**
   a. They sit in the mitochondrial inner membrane.
   b. They pass along electrons received from NADH or FADH$_2$, and as they do, they pump H$^+$ from the matrix into the intermembrane space.
   c. All of the <u>cytochromes</u> (but <u>not</u> NADH dehydrogenase) contain a heme group (i.e., a porphyrin ring containing an atom of iron).
   d. Thirteen of the approximately 100 polypeptides used in the ETC are coded for by mitochondrial DNA. The rest are synthesized from nuclear DNA and imported from the cytoplasm.
   e. Enzymatic defects of the ETC most often occur among those 13 polypeptides that are coded for by mitochondrial DNA. **Why?** → B/c mtDNA has a mutation rate about 10 times that of nuclear DNA.

3. **The ETC—facts you need to know**
   a. **How many different complexes are there on the ETC?** → 5
   b. **NADH drops off its electrons at which complex?** → Complex 1
   c. **What's another name for complex 1?** → NADH dehydrogenase
   d. **Which molecule is the crucial part of complex 1 because it accepts the electrons from NADH?** → FMN
   e. **What are three substances that inhibit complex 1?**
      (1) Amytal (barbiturate)
      (2) Rotenone (insecticide)
      (3) Piericidin A (antibiotic)
   f. **What shuttles electrons between complex 1 and 2?** → Coenzyme Q (a.k.a., ubiquinone)
   g. **FADH$_2$ drops off its electrons at which complex?** → Complex 2
   h. **What is another name for complex 2?** → Cytochrome b
   i. **What are two substances that inhibit complex 2?**
      (1) Antimycin A (antibiotic)
      (2) Malonate
   j. **What shuttles electrons between complex 2 and 3?** → Coenzyme Q (a.k.a., ubiquinone)
   k. **What is another name for complex 3?** → Cytochrome c1
   l. **What shuttles electrons between complex 3 and 4?** → Cytochrome C
   m. **Which complex contains copper (Cu$_{2+}$)?** → Complex 4
   n. **What are other names for complex 4?** → Cytochrome a$^+$a3 or cytochrome oxidase
   o. **We know that oxygen is the terminal electron acceptor in the ETC. At which complex does oxygen act and what does it form as it accepts electrons?** → Oxygen reacts at complex 4; it combines electrons with H$^+$ from the matrix and forms H$_2$O.

p. **Why is the function of oxygen as the terminal electron acceptor so important?** → B/c by removing electrons from complex 4, oxygen allows for continued electron flow down the ETC (Figure 9.3).

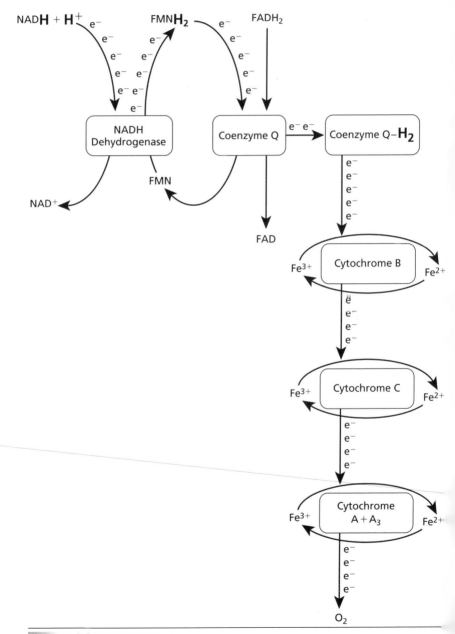

FIGURE 9.3 Electron transport.

q. **Clinical correlations:**
   (1) **Why does lack of oxygen kill someone?** → B/c oxygen is the terminal electron acceptor in the ETC
   (2) **Why does too much oxygen kill someone?** → Cerebral blood stasis; there is no stimulus for the vasculature to move the blood along.
   (3) **Why do we never use 100% oxygen in treating patients?** → 100% easily can catch on fire and blow up the hospital.
   (4) Okay, enough of the tangential info...No oxygen → "backup" of electrons in the ETC → no electron flow down the ETC → no pumping of H^+ into the intermembrane space → no H^+ gradient → no flow of H^+ through ATP synthetase → no ATP synthesis → death
r. **What five substances can inhibit complex 4?**
   (1) Cyanide (CN^-)
   (2) Azide
   (3) Carbon monoxide (CO)
   (4) Hydrogen sulfide (H_2S)
   (5) Lack of oxygen
s. **What's another name for complex 5?** → ATP synthetase
t. **As ATP synthetase uses the flow of H^+ to produce ATP, how many H^+ are required to produce 1 ATP?** → 2 H^+ per ATP produced
u. **What inhibits the function of complex 5 by binding and blocking its H^+ channel?** → Oligomycin

4. **Summary of the names of the five complexes and the two electron carriers** (Table 9.2).

TABLE 9.2 Names of ETC Complexes and Electron Shuttles

Complex 1	NADH dehydrogenase
What shuttles electrons between complex 1 and 2?	Coenzyme Q (a.k.a., ubiquinone)
Complex 2	Cytochrome b
What shuttles electrons between complex 2 and 3?	Coenzyme Q (a.k.a., ubiquinone)
Complex 3	Cytochrome c1
What shuttles electrons between complex 3 and 4?	Cytochrome C
Complex 4	Cytochrome a+a3 (a.k.a., cytochrome oxidase)
Complex 5	ATP synthetase

Do we really care about all these complexes? Let's make this clinically relevant, although these are rare diseases (with a capital R). You will only find these diseases on the boards:

> **HPI:** A 12 yo presents with malaise, weakness, short stature, ataxia, dementia, hearing loss, sensory neuropathy, pigmentary retinopathy, and pyramidal signs, progressing in severity over a period of many years. On muscle biopsy, ragged-red fibers are common. A mild lactic acidosis is diagnosed. After many years of visits to physicians who earned **99th percentile on their boards,** they have decided to come to you. Having read this book, you realize that these physicians have missed a very rare mitochondrial disease. You do some specific mitochondrial assays to determine this is none other than a...?
> → Complex III Deficiency (ubiquinone-cytochrome c oxidoreductase deficiency)
> - Other manifestations include: Infantile histiocytoid cardiomyopathy, fatal infantile encephalomyopathy

Ok, ok! Last one, I promise:

> **HPI:** An 18-month-old child who was normal the first year of life presents with progressive developmental regression, ataxia, lactic acidosis, optic atrophy, ophthalmoplegia, nystagmus, dystonia, pyramidal signs, and respiratory problems. Frequent seizures that have not abated with anticonvulsant therapy have the neurologists confused. They look to you, a **second-year medical student** to make the diagnosis. You say coyly, "Let's order some mitochondrial assays." They are astounded at your brilliance as you make the diagnosis of...?
> → Complex IV deficiency
> - Cytochrome c oxidase deficiency is caused by a defect in Complex IV of the respiratory chain. Myopathies are common with a complex IV deficiency.

5. **The impermeability of the mitochondrial inner membrane:** A problem that must be solved by using four different shuttle systems (the glycerol-3-phosphate shuttle, the malate/aspartate shuttle, the adenine nucleotide carrier, and the phosphate carrier).
   a. **Why can't cytoplasmic NADH enter the mitochondrial matrix in order to drop off its electrons at the ETC?** → B/c the inner membrane is impermeable to it
   b. **How does cytoplasmic NADH get around this barrier?** → It has two options. It can use the malate/aspartate or the glycerol-3-phosphate shuttles.
   c. **Why can't cytoplasmic ADP and Pi enter the mitochondrial matrix to get reformed into ATP?** → B/c the inner membrane is impermeable to them
   d. **How does ADP get around the impermeability of the inner membrane?** → By using the adenine nucleotide carrier (a.k.a., the ADP-ATP translocase)
   e. **How does Pi get around the impermeability of the inner membrane?** → By using the phosphate carrier
   f. **The glycerol-3-phosphate shuttle** (Figure 9.4)
      (1) **Problem:** Cytosolic NADH would like to contribute to the electron transport chain, but it cannot cross the mitochondrial inner membrane.
      (2) **Solution:** The glycerol-3-phosphate shuttle. Essentially, the electrons from cytoplasmic NADH are transferred to $FADH_2$ in the inner membrane. This process has **four steps:**
         (a) NADH reduces DHAP to glycerol-3-P
         (b) G-3-P reacts at the inner membrane to generate $FADH_2$ and regenerate DHAP
         (c) $FADH_2$ takes the H to the electron transport chain
         (d) DHAP leaves the inner membrane and enters the cytoplasm

KEY IDEA: Because $FADH_2$ is used as the final electron carrier in **the gylcerol-3-phosphate shuttle, only two ATP** are produced per cytosolic NADH oxidized.

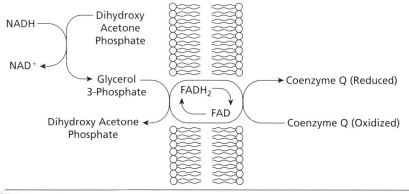

FIGURE 9.4 Glycerol-3-phosphate shuttle.

g. **The malate/aspartate shuttle** (Figure 9.5)
   (1) **Problem:** Cytosolic NADH would like to contribute to the electron transport chain, but it cannot cross the mitochondrial inner membrane
   (2) **Solution:** The malate/aspartate shuttle. Essentially, the electrons from cytoplasmic NADH are transferred to an NADH in the inner membrane. This process has **three steps:**
       (a) Cytoplasmic NADH reduces cytosolic OAA to malate
       (b) Malate reacts at the inner membrane to generate NADH
       (c) NADH takes the H to the electron transport chain

KEY IDEA: Because **NADH** is used as the final electron carrier in the **malate/aspartate shuttle, three ATP are produced** per cytosolic NADH oxidized.

h. **The adenine nucleotide carrier**
   (1) **Problem:** Cytosolic ADP would like to reach ATP synthetase so that it could be reformed into ATP, but it cannot cross the mitochondrial inner membrane.

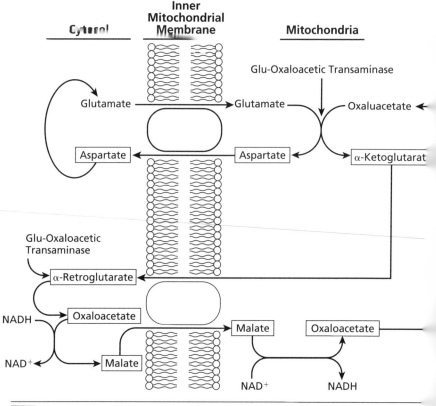

FIGURE 9.5 Malate-aspartate shuttle.

(2) **Solution:** The adenine nucleotide carrier. It is an ADP/ATP exchanger. It transports 1 molecule of ADP from cytosol → matrix; and one molecule of ATP from matrix → cytosol.
(3) **Clinical correlation**
  (a) **Which plant toxin kills you by inhibiting the action of the adenine nucleotide carrier?** → Atractyloside
  (b) **How does atractyloside kill you?** → Inhibition of adenine nucleotide carrier → depletion of intramitochondrial stores of ADP → cessation of ATP synthesis
i. **The phosphate carrier**
  (1) **Problem:** Cytosolic Pi would like to reach ATP synthetase and participate in the creation of ATP, but it cannot cross the mitochondrial inner membrane.
  (2) **Solution:** The phosphate carrier. The phosphate carrier transports Pi from cytosol → matrix.

KEY IDEA: There are three ways that the process of ATP generation using the ETC can be disrupted:
1. **ETC inhibitors:** These inhibit the flow of electrons down the electron transport chain. *exs)* Lack of oxygen, a defective ETC complex, Amytal (barbiturate), Rotenone (insecticide), Piercidin A (antibiotic), Antimycin A (antibiotic), Malonate, cyanide, azide, carbon monoxide, hydrogen sulfide
2. **Uncoupling agents:** These destroy the H^+ gradient in the intermembrane space by making the inner membrane permeable to H^+.
   ex) 2,4-dinitrophenol
   ex) Aspirin in high doses
   - **Why are they called "uncoupling" agents?**
   - **Short answer:** B/c by destroying the H^+, gradient they effectively "uncouple" the work of complexes 1 through 4 from the creation of ATP by complex 5.
   - **Long answer:** The work of complexes 1 through 4 is to pump H^+ into the intermembrane space, thus creating an H^+ gradient. Normally this gradient is coupled to the creation of ATP by having H^+ flow back into the matrix via complex 5 (ATP synthetase). Although complexes 1 through 4 continue to pump H^+, once the inner membrane is made permeable to H^+ it is <u>impossible</u> to create a proton gradient. No proton gradient = no driving force to run ATP synthetase. Essentially, the work of the complexes 1 through 4 has been "uncoupled" from the creation of ATP by complex 5.
3. **ATP synthetase inhibitors:** These inhibit the production of ATP by ATP synthetase. *ex)* Oligomycin
   - **Who cares if ATP synthesis gets disrupted by any of the previously mentioned means?** → B/c lack of ATP = cell damage and/or death.
   - **Which tissues are the first to die from a lack of ATP?** → Brain and nervous tissues. Why? → B/c they require lots of ATP to run the Na^+/K^+ pumps that maintain neuron membrane potentials.
   - **Clinical correlation** → The fact that the brain requires tons of ATP to maintain neuron membrane potentials is the reason why it is the first organ to be adversely affected by oxygen deprivation and the resultant ↓ in ATP synthesis.

6. **The three types of ETC inhibitors and their effects on electron transport, the proton gradient, and ATP synthesis** (Table 9.3).

TABLE 9.3

Type of Inhibitor	Effects
Electron transport inhibitors • Lack of oxygen • Defective ETC complex • Amytal • Rotenone • Antimycin A • Cyanide	• ↓ Electron transport down the ETC • ↓ Proton gradient. Why? B/c ↓ flow of electrons down the ETC = ↓ H^+ pumped into intermembrane space • ↓ ATP synthesis by ATP synthetase. Why? B/c ↓ H^+ gradient = ↓ flux through ATP synthetase
Uncoupling agents (destroy the H^+ gradient) • 2,4-DNP • Aspirin in high doses	• ↓ Proton gradient • ↓ ATP synthesis by ATP synthetase. Why? B/c ↓ H^+ gradient = ↓ flux through ATP synthetase • Electron transport is unaffected. Why? B/c electrons still flow down the ETC and cause H^+ to be pumped into the intermembrane space. The only difference is that now the H^+ does not stay there.
ATP synthetase inhibitors • Oligomycin	• ↓ ATP synthesis by ATP synthetase • ↑ Proton gradient. Why? B/c inhibition of ATP synthetase causes a backup of H^+ into the intermembrane space • ↓ Electron transport down the ETC. Why? B/c the ↑ $[H^+]$ in the intermembrane space creates a gradient so strong that complexes 1-4 c are no longer able to pump H^+ against it. As a result, these complexes stop working

7. **Specific inhibitors of oxidative phosphorylation categorized by mechanism** (Table 9.4)

TABLE 9.4

Mechanisms	Inhibitors
Inability of electrons to flow through a certain part of the ETC	• Defective ETC complex (usually genetic cause; usually from mtDNA)
Attaches to H^+ and brings it across the inner membrane, thus destroying the H^+ gradient and inhibiting ATP synthesis	• 2,4-dinitrophenol
Inhibits complex 1	• Amytal (barbiturate) • Rotenone (insecticide) • Piercidin A (antibiotic)
Inhibits complex 2	• Antimycin A (antibiotic) • Malonate
Inhibits complex 4	• Cyanide (CN^-) • Azide • Carbon monoxide (CO) • Hydrogen sulfide (H_2S)
Inhibits the removal of electrons from complex 4	• Lack of oxygen
Binds and blocks the H^+ channel on ATP synthetase	• Oligomycin
Inhibits the adenine nucleotide carrier that exchanges ATP and ADP across the inner membrane	• Atractyloside (plant-derived toxin)

Chapter 10
Carbohydrate Metabolism

A. An Overview of Glycogen Synthesis and Regulation
1. Why is glycogen necessary if gluconeogenesis can occur in the liver and kidneys? → B/c gluconeogenesis is slow to respond to a drop in blood glucose. In contrast, glycogen is rapidly metabolized.
2. What are the two main places that glycogen is <u>synthesized and stored</u>?
   a. Liver
   b. Skeletal muscle
3. What is the rate-limiting enzyme and reaction of glycogen <u>synthesis</u>?

$$\text{UDP-glucose} \xrightarrow{\text{glycogen synthase}} \text{glycogen}$$

<u>Key idea</u>: There are two pathways for glycogen <u>degradation</u>—the normal one and the weird one.
1. The normal one occurs in the <u>cytoplasm</u> and involves the <u>majority</u> of glycogen.
2. The weird one occurs in the <u>lysosomes</u> and involves <u>only small amounts</u> of glycogen.

4. **The normal pathway:** Glycogen → glucose-1-P → glucose-6-P → glucose.
   a. What is the rate-limiting enzyme and reaction?

$$\text{glycogen} \xrightarrow{\text{glycogen phosphorylase}} \text{glucose-1-P}$$

b. **Why is it that the <u>liver can</u> degrade glycogen and release that glucose into the blood, but <u>muscles cannot</u>?** → B/c muscles do not have the enzyme glucose-6-phosphatase. Therefore they cannot convert glucose-6-P (which cannot leave the cell) into free glucose (which can).
5. **The "weird" pathway:** Degradation of glycogen in lysosomes. It is considered "weird" because the purpose of such degradation is unknown.

KEY IDEA: Both glycogen synthesis and degradation are always occurring. It is the difference between the rates of these reactions that determines whether glycogen is stored or degraded.

6. **In the LIVER, there are four things that <u>promote</u> net glycogen <u>synthesis</u>.**
   a. **Glucose-6-phosphate** (stimulates synthase; inhibits phosphorylase).
   b. **Insulin** (activates synthase (via dephosphorylation); deactivates phosphorylase (via dephosphorylation)
   c. **Glucose** (inhibits phosphorylase)
   d. **ATP** (inhibits phosphorylase)
7. **In the LIVER, there are three things that <u>promote</u> net glycogen degradation.**
   a. **Glucagon** (activates phosphorylase (via phosphorylation); inactivates synthase (via phosphorylation)
   b. **Epinephrine** (same mechanism as glucagon)
   c. **Caffeine** (same mechanism as glucagon)
8. **Flashbacks:**
   a. Remember the **mechanism of action of glucagon and epinephrine:** Glucagon or epinephrine binds to G-protein linked receptor → ↑ cAMP → ↑ protein kinase A → phosphorylation
   b. Remember **the mechanism of action of caffeine:** Caffeine inhibits phosphodiesterase → ↑ cAMP → ↑ protein kinase A → phosphorylation
9. **In SKELETAL MUSCLE, there are three things that <u>promote</u> net glycogen <u>synthesis</u>.**
   a. **Glucose-6-phosphate** (stimulates synthase; inhibits phosphorylase)
   b. **Insulin** (activates synthase (via dephosphorylation); inactivates phosphorylase (via dephosphorylation)
   c. **ATP** (inhibits phosphorylase)
10. **Flashback** → Remember the **mechanism of action of insulin:** Insulin binds to tyrosine kinase receptor → ↓ cAMP → DE-phosphorylation
11. **In SKELETAL MUSCLE, there are five things that <u>promote</u> net glycogen degradation.**
    a. **Glucagon** (activates phosphorylase (via phosphorylation); inactivates synthase (via phosphorylation)
    b. **Epinephrine** (same mechanism as glucagon)
    c. **Caffeine** (same mechanism of glucagon)
    d. **Ca^{++}** (binds to phosphorylase and stimulates it)
    e. **AMP** (binds to inactive phosphorylase and activates it)

B. GLYCOGEN STORAGE DISEASES

> **KEY IDEA:** All glycogen storage diseases result in an accumulation of glycogen within cells.

Some clinically important glycogen problems:

HPI: 5-month-old child to **Jewish** parents presents with failure to thrive. On PE, **hepatomegaly** (enlarged liver) and several **petechiae** are noted. Height and weight are at the 2nd percentile for age. Laboratory testing reveals **low blood sugar** levels and an increased amount of certain fats and acids in the blood, which give the diagnosis. There is a characteristic **"doll-like"** facial appearance and a short, **"stocky"** morphology in the child resulting from an atypical distribution of fat tissue in the body. What is this poor child suffering from?
→ Von Giercke's disease (Type 1 glycogen storage disease)
- **Problem:** Cannot convert G-6-P into glucose. **Why not?** → **Glucose-6-phosphatase deficiency** in liver, kidneys, and intestines.

HPI: 12 yo presents with chronic decreased **stamina and fatigue** with minimal exercise. Additionally, he has significant muscle pain and stiffness, which **abates with cessation** of exercise. His desire to play basketball has resulted in prolonged (1 to 2 days) muscle pain and swelling during tryouts, which force him to quit every season. There was one episode last year when he was diagnosed with **acute renal failure** after a particularly difficult tryout. His parents have noted that his muscles appear to have atrophied over the past few years. He is **mentally sharp** and is in the top 5% of his class. What is the diagnosis?
→ McArdle's syndrome (type 5 glycogen storage disease)
- **Problem:** Cannot degrade intramuscular glycogen stores. **Why not?** → **Skeletal muscle glycogen phosphorylase deficiency.**

HPI: 3-month-old child presents with weakness and **hypotonia**, which have included breathing and feeding difficulties. The birth and family history were noncontributory. On PE, **tachypnea, cardiomegaly, macroglossia, hepatomegaly,** and increased muscle bulk with hypotonia are noted. **Cardiac failure from LV enlargement** and outflow obstruction have lead to respiratory and feeding difficulties by echocardiography. The **serum CK and AST** are elevated, (suggesting a muscle and liver pathology). A **skin biopsy** provides the definitive diagnosis of...?
→ Pompe's disease (type 2 glycogen storage disease)
- **Problem:** Lysosomal degradation of glycogen does not work. **Why not?** → Alpha-1,4 glucosidase (a.k.a., acid maltase) deficiency. The skin biopsy tests for alpha-1,4 glucosidase activity in cultured fibroblasts.

> **HPI:** 4 yo child presents with a swollen abdomen. Further examination determines this to be **hepatomegaly**. He is described as being weaker than his 3 yo brother, which is pronounced when they wrestle. On PE, height and weight are delayed for age. Laboratory work shows perpetually **low blood sugar** levels on fasting. The EKG is **normal**. A diagnosis is made and a high-protein diet and prevention of prolonged periods of fasting helps the patient significantly. By 14 yrs of age, the child is completely **normal**. What is this?
> → Cori's or Forbes disease (type 3 glycogen storage disease)
> - **Problem:** Glycogen degradation is partially impaired. **Why?** → Deficiency of the debranching enzyme alpha-1,6-glucosidase

C. Metabolism of the monosaccharides fructose and galactose
1. **General principles**
   a. **What are the three monosaccharides?**
      (1) Glucose
      (2) Fructose
      (3) Galactose
   b. **What are the three disaccharides?**
      (1) Lactose (milk sugar)   [lactose = glucose + galactose]
      (2) Sucrose (cane sugar)   [sucrose = glucose + fructose]
      (3) Maltose   [maltose = glucose + glucose]
2. **Fructose**
   a. **What is the major source of fructose?** → From sucrose (table sugar or cane sugar)
   b. **Does the presence of fructose in the blood elicit release of insulin?** → Not really—it causes release of only a very small amount of insulin.
   c. **Where does most metabolism of fructose happen?** → In the liver
   d. **Is entry of fructose into a cell affected by insulin levels?** → No—entry of fructose into cells is <u>independent</u> of insulin
   e. **What is the fate of fructose once it enters a cell?** → Enters glycolysis by being converted to the glycolytic intermediates DHAP or glyceraldehyde 3 phosphate
3. **There are really only two enzymes and reactions that you need to know in fructose metabolism:**
   a. **Fructokinase.** As soon as fructose enters the cell it gets phosphorylated by fructokinase to create fructose–1-P.

$$\text{fructose} \xrightarrow[\text{(fast reaction)}]{\text{fructokinase}} \text{fructose–1-P}$$

> **HPI**: 40 yo man was in USOH (usual state of health) when he presented. He has **no complaints** and does not know why he is at the physician's office. The overzealous physician has prescribed every test available for a physical (not under managed care) and it appears the patient has **excess fructose in the urine**. What is the diagnosis?
> → Essential fructosuria
> - Mutation in the fructokinase gene → cells cannot metabolize fructose → fructose accumulates in cells and ultimately appears in the urine
> - Completely benign and asymptomatic

  b. **Aldolase B.** Fructose–1–P is cleaved by aldolase B to create DHAP or glyceraldehyde.

> $$\text{fructose–1–P} \xrightarrow[\text{(slow reaction)}]{\text{Aldolase B}} \text{DHAP or glyceraldehyde}$$

> **HPI**: 3-month-old infant presents with severe **hypoglycemia**. Clinical symptoms include severe abdominal tenderness, vomiting, and malaise following ingestion of certain sweet foods. On PE, **hepatomegaly** is noted. BUN/Cr levels are elevated (**kidney** dysfunction). A diagnosis is made and the parents feed the child sucrose and fructose-free food, which ameliorate the condition. What was the child suffering from?
> → Hereditary fructose intolerance
> - **Lack of aldolase B** → cells cannot metabolize fructose–1–P → fructose–1–P accumulates in cells → severe hypoglycemia

4. **Galactose**
    a. **What is the major source of galactose?** → Lactose (milk sugar)
    b. **Does the presence of galactose in the blood elicit release of insulin?** → No.
    c. **Is entry of galactose into a cell affected by insulin levels?** → No—entry of galactose into cells is independent of insulin.
    d. **What are the three possible fates of galactose once it enters a cell?**
        (1) Enters glycolysis (all cells) by being converted to glucose-6-phosphate via the following pathway: Galactose → UDP-galactose → UDP-glucose → glucose-1-P → glucose-6-P.
        (2) Enters gluconeogenesis (liver and kidney cells only).
        (3) Structural precursor in the synthesis of glycolipids, glycoproteins, and glycosaminoglycans.

D. GLUCOSE METABOLISM: UP CLOSE AND PERSONAL
 1. **There is glucose in your blood. How did it get there?** (three sources)
    a. It came **from the diet** and was absorbed via the small intestine.
    b. It came from **gluconeogenesis** in the liver or kidneys.
    c. It came from **degradation of liver glycogen**.
 2. **Nutritional correlation:** → **If you do not get enough carbohydrates in your diet, what are the metabolic effects?**
    a. Glycogen stores depleted
    b. ↑ Degradation of muscle proteins (for use as energy)
    c. ↑ Breakdown of TGs; ↓ TG synthesis
    d. ↑ Beta oxidation of FFAs; ↓ de novo synthesis of FFAs
    e. ↑ Ketone production (in order to produce energy for the CNS)
 3. **What stimulates glucose uptake into skeletal muscle and fat cells?**
    a. Insulin
    b. **How?** → It increases the number and activity of Glut 4 transport proteins in skeletal muscle and in adipose tissue.

> **KEY IDEA:** Glucose uptake into the following three tissues is NOT affected by insulin levels:
> 1. Brain
> 2. Liver
> 3. RBCs

 4. **What three things inhibit glucose uptake in skeletal muscle and fat cells?**
    a. Cortisol
    b. HGH
    c. High blood [FAA]
 5. **What are the two fates of glucose once it enters a cell?**
    a. Enters glycolysis **(or)**
    b. Is stored as glycogen

> **MASSIVE KEY IDEA:** Glucose and fatty acids are, in a sense, competing energy substrates. Their metabolism runs in opposite directions!

 6. ↑ **Fat consumption** = ↓ **use of carbohydrates for energy and** ↑ **carbohydrate storage.** (The opposite is also true: ↓ fat consumption = ↑ use of carbohydrates for energy and ↓ carbohydrate storage.)
    a. Part 1: "↑ fat consumption = ↓ use of carbohydrates for energy"
       - **How?**
         (1) High blood [FFA] inhibits glucose uptake into cells
         (2) High blood [FFA] decreases flux through glycolysis
    b. Part 2: "↑ fat consumption = ↑ carbohydrate storage"
       - **How?**
         (1) High blood [FFA] stimulates gluconeogenesis
 7. ↑ **Carbohydrate consumption** = ↓ **use of fat for energy and** ↑ **fat storage.** (The opposite is also true: ↓ carbohydrate consumption = ↑ use

of fat for energy **and** ↓ fat storage.) **Nutritional correlation:** → This is a biochemical rationale used to promote the idea of low-carbohydrate diets! ↓ carbohydrate intake → ↑ use of fat for energy and ↓ fat storage → ↓ "lovehandles" (For a full discussion of the merits and disadvantages of a low-carbohydrate diet, see Chapter 14: Review of Metabolism.)
  a. Part 1: "↑ carbohydrate consumption = ↓ use of fat for energy"
  - How?
    (1) High blood (glucose) → ↑ flux through glycolysis → ↑ acetyl CoA → ↓ **beta oxidation of FFA**
  b. Part 2: "↑ carbohydrate consumption = ↑ fat storage"
  - How?
    (1) **By ↑ [acetyl CoA].** High blood [glucose] → ↑ flux through glycolysis → ↑ acetyl CoA → ↑ de novo FFA synthesis → ↑ TG synthesis and ↓ TG breakdown
    (2) **By ↑ [insulin].** High blood glucose → ↑ insulin → ↑ TG synthesis and ↓ TG breakdown
    (3) **By ↑ levels of two important enzymes involved in de novo FFA synthesis.** Prolonged periods of high blood glucose → ↑ acetyl CoA carboxylase and ↑ fatty acid synthase → ↑ fatty acid synthesis → ↑ TG synthesis

---

<u>KEY IDEA</u>: There is only one hormone that has the effect of ↓ blood glucose. Know it. → <u>Insulin</u>.
- How does insulin ↓ blood glucose?
  1. By ↑ **glucose uptake** into muscle and fat cells (by recruiting glut 4 receptors)
  2. By ↓ **glycogen breakdown** and ↑ **glycogen synthesis** in liver and muscle
  3. By ↓ **gluconeogenesis** in liver and kidney

<u>KEY IDEA</u>: **These four hormones all have the effect of ↑ blood glucose. Memorize them** (Table 10.1).

TABLE 10.1 Four Hormones that Raise Blood Glucose Levels

Hormones	How Does It ↑ Blood Glucose?
Glucagon from pancreas alpha cells	(1) ↑ Glycogen breakdown and ↓ glycogen synthesis in liver (but NOT in muscle) (2) ↑ Gluconeogenesis in liver and kidney
Cortisol from adrenal cortex	(1) Enhances the effect of glucagon in degrading glycogen (2) ↑ Gluconeogenesis (3) ↓ Glucose uptake into muscle and fat cells
Epinephrine and norepinephrine from adrenal medulla	(1) ↑ Glycogen breakdown (2) ↑ Gluconeogenesis
HGH from anterior pituitary	(1) ↑ Gluconeogenesis (2) ↓ Glucose uptake into muscle and fat cells

8. **Regulation of glucose homeostasis**
   a. **Scenario #1: "Help, I'm <u>running low on glucose</u>!" (That's your body talking.) If blood glucose levels fall too low (e.g., prolonged fast), why is this bad?**
      (1) B/c glucose is the <u>preferred fuel for all tissues</u> (except heart muscle and renal cortex, which prefer ketone bodies over glucose).
      (2) B/c glucose is the <u>only fuel for RBCs</u>.
      (3) B/c glucose is <u>one of only two fuels for the CNS</u> (ketones are the other).
      (4) Moderate lack of blood glucose = fatigue and inability to concentrate.
      (5) Severe lack of glucose = fainting, coma and/or death.
   b. **If blood glucose levels fall <u>too low</u>, how is it sensed?** → By the alpha cells of the pancreas
   c. **If blood glucose levels fall <u>too low</u>, what is the body's response?**
      (1) **Release of the four hormones** that ↑ blood glucose levels
         (a) **Glucagon** from pancreas alpha cells
         (b) **Cortisol** from the adrenal cortex
         (c) **Epinephrine and norepinephrine** from the adrenal medulla
         (d) **HGH** from the anterior pituitary
      (2) **The liver does four things:**
         (a) ↑ **Gluconeogenesis** (This releases glucose into the blood.)
         (b) ↑ **Glycogen breakdown** (This releases glucose into the blood.)
         (c) ↑ **Production of ketone bodies** (This provides an alternative fuel source besides glucose.)
         (d) ↑ **Conversion of fatty acids into acetyl CoA (This occurs via beta oxidation.)**
            - Why does this make sense? → B/c the ↑ in acetyl CoA derived from <u>fats decreases the need to use glucose</u> to generate acetyl CoA. Therefore glucose is "spared" from being used as a substrate for generating acetyl CoA in the liver, and more glucose is available for release into the bloodstream.
   d. **Scenario #2: "Por favor, ayudame. Tengo demasiada glucosa en mi sangre." (That's your body speaking Spanish and telling you that your <u>blood glucose levels are TOO HIGH</u>.)**
      (1) **If blood glucose levels rise too high (e.g., diabetes), why is this bad?**
         (a) It causes <u>hyperosmolarity of the blood</u>. The extra osmoles attract water, accounting for the polyuria, thirst, and dehydration seen in diabetes.
         (b) It is thought to <u>contribute to the chronic effects of diabetes</u> (e.g. premature atherosclerosis, retinopathy, nephropathy, neuropathy by altering membrane proteins through glycosylation of free amino groups.
      (2) **If blood glucose levels rise <u>too high</u>, how is it sensed?** → By the beta cells of the pancreas
      (3) **If blood glucose levels rise <u>too high</u>, what is the body's response?**

(a) Beta cells of pancreas **release insulin**
(b) **Excess glucose is filtered into the kidney** for excretion in the urine
(c) **The liver does three things:**
   (i) ↑ **Glycogen synthesis** (This lowers blood glucose by storing it.)
   (ii) ↓ **Gluconeogenesis** (This stops the production of new glucose.)
   (iii) ↓ **Ketone body production** (Why waste energy producing ketones when there is tons of glucose around?)
9. **Flux through glycolysis and gluconeogenesis is influenced by the carbohydrate content of the diet, and is mediated by insulin and glucagon.**
   a. High [carbohydrate] = ↑ blood glucose = ↑ insulin = ↑ flux through glycolysis, ↓ flux through gluconeogenesis
   b. Low [carbohydrate] = ↓ blood glucose = ↑ glucagon = ↓ flux through glycolysis, ↑ flux through gluconeogenesis

E. GLYCOLYSIS: USING MONOSACCHARIDES TO PRODUCE ENERGY AND IMPORTANT METABOLIC SUBSTRATES
1. **Overview of key ideas**
   a. Glycolysis occurs in the cytoplasm of all cells.
   b. Glycolysis can occur under anaerobic or aerobic conditions.
   c. The three most important enzymes in glycolysis are hexokinase/glucokinase, PFK-1, and pyruvate kinase because they represent the three points of regulation.
2. **Metabolic context:** Glycolysis is always occurring. Flux is increased during the postabsorptive state (mediated by insulin). Flux is decreased during times of low blood sugar (mediated by glucagon) or stress (mediated by cortisol and epinephrine).
3. **There are two types of glycolysis: Aerobic and anaerobic**
   a. If there is enough O_2 around, then aerobic glycolysis occurs: 1 glucose → 2 pyruvate
   • In addition to pyruvate, 2 NADH and 2 ATP are generated per molecule of glucose.
   b. If there is NOT enough O_2 around, then anaerobic glycolysis occurs: 1 glucose → 2 pyruvate → 2 lactate
   • In addition to lactate, 2 ATP are produced per molecule of glucose.
   • Note that under anaerobic conditions, there is NO net generation of NADH.
   c. **Clinical correlation**
      (1) **Why is it that patients with impaired circulatory function (e.g., from a myocardial infarction, pulmonary embolism, or hemorrhagic shock) will often have elevated blood lactate levels (a.k.a., "lactic acidosis")?**
      → B/c of ↑ anaerobic glycolysis as a result of poor oxygen delivery to tissues.

- Here is the pathway: Poor circulation → ↓ oxygen delivery to tissues → ↑ anaerobic glycolysis → ↑ production of lactate
4. **Starting products, end-products, and energy generation in glycolysis**
   a. **Aerobic glycolysis:** 1 glucose → 2 pyruvate + 2 ATP + 2 NADH
   b. **Anaerobic glycolysis:** 1 glucose → 2 lactate + 2 ATP
   c. **What is the total number of ATP generated, both directly and indirectly, by aerobic glycolysis?** → There is controversy on this question.
      (1) Some books say **8 ATP** (direct production of 2 ATP; and indirect production of **3 ATP** per NADH).
      (2) Other books say **6 ATP** (direct production of 2 ATP; and indirect production of only **2 ATP** per NADH)
      (3) **Why do some books say only 2 ATP per cytoplasmic NADH?** → B/c of the energy loss involved in transporting electrons from cytoplasmic NADH into the mitochondria (to reach the electron transport chain)
   d. **What is the total number of ATP generated, both directly and indirectly, by anaerobic glycolysis?** → 2 (Direct production of 2 ATP. No production of NADH)
   e. **What are the five main reactants that enter glycolysis and where do they come from?**
      (1) **Glucose** (absorbed from the blood or produced by gluconeogenesis in liver and kidneys)
      (2) **Glucose-6-P** (from the breakdown of glycogen [in liver and skeletal muscles only])
      (3) **Fructose** (absorbed from the blood)
      (4) **Galactose** (absorbed from the blood)
      (5) **Glycerol phosphate** (from degradation of TGs)

KEY IDEA: Substrates cannot enter glycolysis unless they are phosphorylated!
exs) Glucose must first be phosphorylated by either glucokinase (liver) or hexokinase (all other cells); galactose must first be phosphorylated by galactokinase; fructose must first be phosphorylated by fructokinase.

---

   f. **What are the two end-products of glycolysis and where do they go?**
      (1) **Pyruvate** (This is the end-product under aerobic conditions.) Pyruvate has three fates in nonhepatic cells and four fates in hepatic cells. They are discussed later in this chapter.
      OR
      (2) **Lactate** (This is the end-product under anaerobic conditions.)
         - Lactate has three fates:
           (a) **Is built up** within the cell
           (b) Is sent to the liver or kidneys to be a **substrate for gluconeogenesis** (This is the first part of the "Cori cycle.")
           (c) **Is sent to the liver and heart muscle** for entry into the TCA cycle (after first being reconverted to pyruvate)

> **KEY IDEA:** Glycolysis requires NAD$^+$. This means that for glycolysis to occur, cytosolic NAD$^+$ **must be regenerated.** How is this accomplished?
> 1. (Under **low O$_2$** conditions) NAD$^+$ is regenerated in the **conversion of pyruvate to lactate.**
> 2. (Under **high O$_2$** conditions) NAD$^+$ is regenerated when **NADH "drops off" its electrons at the mitochondria** (using the malate/aspartate shuttle and glycerol-3 phosphate shuttles to take its electrons to the ETC).

    g. Why is the conversion of pyruvate to lactate by lactate dehydrogenase the <u>crucial reaction</u> that allows RBCs, and other cells with few or no mitochondria, to survive? → Since they have no mitochondria (and therefore no TCA cycle), these cells rely on glycolysis as their only means of generating ATP! However, for glycolysis to occur, NAD$^+$ must be regenerated. How can these cells regenerate NAD$^+$? Since they do not have electron transport chains (remember, no mitochondria = no ETC), **their main means of regenerating NAD$^+$ is via the conversion of pyruvate to lactate.**

5. **Three key glycolytic enzymes and the reactions they catalyze.** The three most important enzymes in glycolysis are hexokinase/glucokinase, PFK-1, and pyruvate kinase because they represent the three points of regulation.
   a. **Hexokinase or glucokinase:** Phosphorylates glucose to create glucose-6-P

$$\text{glucose} \xrightarrow{\textit{hexokinase or glucokinase}} \text{glucose-6-P}$$

    (1) Which enzyme has a <u>high affinity</u> for glucose (i.e., low Km), works <u>slowly</u> (i.e., low Vmax), is <u>feedback inhibited</u> by glucose-6-P, and is found <u>in all cells besides the liver</u>? → Hexokinase.
    (2) Which enzyme has a <u>low affinity</u> for glucose (high Km), works very <u>quickly</u> (high Vmax), is <u>not feedback inhibited</u> by glucose-6-P, and is found <u>only in the liver</u>? → Glucokinase.
   b. **Phosphofructokinase 1 (PFK-1):** phosphorylates fructose-6-phosphate

$$\text{fructose-6-P} \xrightarrow{\textit{PFK-1}} \text{fructose-1,6-BP}$$

> **KEY IDEA:** PFK-1 is the **rate limiting step of glycolysis!** In other words, glycolysis only proceeds as fast or as slow as PFK-1 catalyzes this reaction.

    (1) **What <u>increases</u> flux through glycolysis by <u>stimulating</u> PFK-1?**
      (a) Direct
        • Fructose-2,6-bisphosphate (F26BP)
        • AMP or ADP

(b) <u>Indirect</u>
  • Insulin (via ↑ production of F26BP)
(2) What <u>decreases</u> flux through glycolysis by <u>inhibiting</u> PFK-1?
  (a) <u>Direct</u>
    • Citrate
    • ATP
  (b) <u>Indirect</u> (all of them reduce stimulation of PFK-1 via ↓ production of F26BP)
    • Glucagon
    • Epinephrine
    • Cortisol
    • Caffeine
c. **Pyruvate kinase:** Converts phosphoenolpyruvate (PEP) to pyruvate

(1) What <u>increases</u> flux through glycolysis by <u>stimulating</u> **pyruvate kinase?**
  (a) <u>Direct</u> **Fructose-1,6-BP** (in liver only)
    • **Note:** The fact that F16BP stimulates pyruvate kinase is an example of "feedforward" regulation. Basically, PFK-1 (an upstream reaction) produces F16BP, which then goes and stimulates pyruvate kinase.
  (b) <u>Indirect</u>. **Insulin** (reduces inhibition of pyruvate kinase via ↓ production of protein kinase A)
(2) What <u>decreases</u> flux through glycolysis by <u>inhibiting</u> **pyruvate kinase?**
  (a) <u>Direct</u>
    • **Protein kinase** A (in liver)
    • **Acetyl CoA**
    • **ATP**
    • **Alanine**
  (b) <u>Indirect</u> (all of these work via ↑ production of protein kinase A, except for FFA)
    • **Glucagon**
    • **Epinephrine**
    • **Cortisol**
    • **Caffeine**
    • **FFA** (via ↑ acetyl CoA resulting from beta oxidation)
6. **Summary of the regulation of glycolysis**
  a. **What is the main <u>stimulator</u> of glycolysis?** → F26 bisphosphate
  b. **How does it stimulate glycolysis?** → By allosteric stimulation of PFK-1
  c. **How is F26BP regulated?**
    (1) Insulin causes ↑ production of F26BP (thus stimulating glycolysis).
    (2) Glucagon causes ↓ production of F26BP (thus reducing stimulation of glycolysis).

d. **What are all the different substances that increase flux through glycolysis?**
   (1) Direct stimulators
       (a) **F26BP** (stimulates PFK-1) (**Note:** F26BP is a molecule that represents a key means of hormonal control over glycolysis.)
       (b) **AMP/ADP** (stimulates PFK-1). (Why does this make sense? → AMP/ADP signals low energy conditions.)
       (c) **F16BP** (stimulates pyruvate kinase). (Why does this make sense? → F16BP signals the buildup of an intermediate product.)
   (2) Indirect stimulators. **Insulin** (stimulates glycolysis in many different ways, to be discussed shortly)
e. **What are all the different substances that decrease flux through glycolysis?**
   (1) Direct inhibitors
       (a) **Glucose-6-P** (inhibits hexokinase). (Why does this make sense? → G6P signals a buildup of starting products.
       (b) **Citrate** (inhibits PFK-1). (Why does this make sense? → Citrate represents the buildup of a TCA intermediate.)
       (c) **ATP** (inhibits PFK-1 and pyruvate kinase). (Why does this make sense? → ATP signals high-energy conditions.)
       (d) **Acetyl CoA** (inhibits pyruvate kinase). (Why does this make sense? → Acetyl CoA represents the buildup of an end-product of glycolysis.)
       (e) **Protein kinase A** (inhibits pyruvate kinase). (Note: Protein kinase A is a molecule that represents a key means of hormonal control over glycolysis.)
       (f) **Alanine** (inhibits pyruvate kinase)
   (2) Indirect inhibitors
       (a) **Glucagon**
           • How?
           • Reduces stimulation of PFK-1 (via ↓ production of F26BP)
           • Inhibits pyruvate kinase (via creation of protein kinase A)
           • ↓ Transcription in the liver of hexokinase, PFK-1, and pyruvate kinase
       (b) **Epinephrine and norepinephrine**
           • How?
           • Reduces stimulation of PFK-1 (via ↓ production of F26BP)
           • Inhibits pyruvate kinase (via creation of protein kinase A)
       (c) **Cortisol** (How? The same two ways as does epinephrine)
       (d) **Caffeine** (How? The same two ways as does epinephrine)
       (e) **FFA** (How? Via ↑ [acetyl CoA] created from beta oxidation)
f. **Spotlight on insulin: What are the three ways that insulin increases flux through glycolysis?**
   (1) Increases synthesis of three glycolytic enzymes in the liver:
       (a) Glucokinase
       (b) PFK-1
       (c) Pyruvate kinase

(2) Stimulates PFK-1 (via ↑ production of F26BP)
  (3) Reduces inhibition of pyruvate kinase in the liver (via ↓ production of protein kinase A)
g. **Spotlight on glucagon: What are the three ways that glucagon decreases flux through glycolysis?**
  (1) Decreases synthesis of three glycolytic enzymes in the liver
   (a) Glucokinase
   (b) PFK-1
   (c) Pyruvate kinase
  (2) Reduces stimulation of PFK-1 (via ↓ production of F26BP)
  (3) Inhibits pyruvate kinase in the liver (via ↑ production of protein kinase A)
7. Once aerobic glycolysis is finished, **pyruvate must still be converted to acetyl CoA** to enter the TCA cycle. This crucial reaction is catalyzed by **pyruvate dehydrogenase complex** and is a key point of metabolic regulation.
 a. **Once aerobic glycolysis produces pyruvate, what happens to it?** → Pyruvate enters into the mitochondrial matrix.
 b. **What happens to pyruvate once it is in the mitochondrial matrix?** → It is decarboxylated by "pyruvate dehydrogenase complex" to yield acetyl CoA.
 c. **What are the five coenzymes required by pyruvate dehydrogenase complex to catalyze this reaction?**
  (1) TPP (a.k.a., **t**hiamin, a.k.a., vitamin B_1)
  (2) FAD (a.k.a., **r**iboflavin, a.k.a., vitamin B_2)
  (3) NAD (a.k.a., **n**iacin, a.k.a., vitamin B_3)
  (4) CoA (a.k.a., **p**antothenate, a.k.a., vitamin B_5)
  (5) Lipoic acid
 d. **Nutritional correlation** → Notice that four of the five coenzymes required for this reaction are B vitamins! This is one reason that deficiencies of B vitamins cause inhibition of cellular energy production. Observe the pathway: Lack of B vitamins → ↓ function of pyruvate dehydrogenase → ↓ flux through TCA → ↓ energy production
 e. **What one thing increases production of acetyl CoA by stimulating pyruvate dehydrogenase complex?** → ADP (causes dephosphorylation—and thus stimulation—of pyruvate dehydrogenase)
 f. **What factors decrease production of acetyl CoA by inhibiting pyruvate dehydrogenase complex?**
  (1) Direct inhibitors
   (a) **Acetyl CoA (this is the most potent inhibitor)**
   (b) **ATP** (causes phosphorylation—and thus inhibition—of pyruvate dehydrogenase)
   (c) **NADH**
  (2) Indirect inhibitors (via production of acetyl CoA)
   (a) **Fatty acids** (produce acetyl CoA via beta oxidation)
   (b) **Ketones** (get degraded to yield acetyl CoA)

F. GLUCONEOGENESIS: USING ENERGY TO CREATE GLUCOSE IN THE LIVER AND KIDNEYS

1. **Overview of key ideas**
   a. Gluconeogenesis is a crucial reaction because it creates glucose.
   b. Gluconeogenesis requires energy input. Therefore low-energy conditions (i.e., ↑ AMP) inhibit gluconeogenesis, whereas high-energy conditions (i.e., ↑ ATP) stimulate it.
   c. Ninety percent of the total amount of glucose created by gluconeogenesis comes from the liver; 10% comes from the cortex of the kidneys. (During a prolonged fast, the contribution from the kidneys increases.)
2. **Location:** Partly in the mitochondrial matrix, and partly in the cytoplasm, of liver and kidney cells.
3. **Reaction summary:** 1 pyruvate (or 1 OAA) + 6 ATP equivalents→ 1 glucose

---

HUGE KEY IDEA: What are the substances that can enter gluconeogenesis and where do they come from?
1. These two substrates enter gluconeogenesis DIRECTLY
   a. **Pyruvate**
      (1) Comes from aerobic glycolysis, **or**
      (2) Deamination of certain gluconeogenic amino acids, **or**
      (3) The conversion of lactate to pyruvate
   b. **Oxaloacetate.** Comes from TCA cycle; may have come ultimately from deamination of certain gluconeogenic amino acids
2. These four substrates enter gluconeogenesis INDIRECTLY
   a. **Lactate**
      (1) Where does it come from? → Glycolysis under anaerobic conditions
      (2) How does it enter gluconeogenesis? → Gets converted to pyruvate
   b. **Glycerol**
      (1) Where does it come from? → Degradation of TGs
      (2) How does it enter gluconeogenesis? → Gets converted to glyceraldehyde-3-P
   c. **Glucogenic amino acids (e.g., alanine)**
      (1) Where do they come from? → From the diet **or** from deamination of muscle proteins
      (2) How do they enter gluconeogenesis? → They get converted to pyruvate, OAA, or alpha-KG
      (3) But how does alpha-KG enter gluconeogenesis? → It gets converted to OAA in the TCA cycle
   d. **Propionyl CoA**
      (1) Where does it come from? → From beta oxidation of odd-numbered fatty acids or from deamination of certain glucogenic amino acids
      (2) How does it enter gluconeogenesis? → It gets converted ultimately to succinyl CoA, which is a TCA cycle intermediate that can yield OAA.

KEY IDEA: Notice that acetyl CoA is NOT one of the substances that can enter gluconeogenesis. This is the reason that fat cannot be converted into glucose!

KEY IDEA: The first two reactions of gluconeogenesis occur in the mitochondrial matrix. The rest occur in the cytoplasm.
- Wait a second. If gluconeogenesis begins in the mitochondrial matrix, but ends in the cytoplasm, how does the intermediate structure (OAA) get from one place to the other? → **The malate shuttle...**
  1. In the mitochondrial matrix
     a. OAA is converted to malate.
     b. Malate exits into the cytoplasm.
  2. In the cytoplasm
     a. Malate is converted back to OAA.

4. **The glucose that gets produced by gluconeogenesis—what are its three possible fates?**
   a. Released into the blood
   b. Stored as glycogen
   c. Rarely shunted into glycolysis
5. **The four key gluconeogenic enzymes and the reactions they catalyze.**
   - The four most important enzymes in gluconeogenesis are pyruvate carboxylase, PEPCK, fructose-1,6-bisphosphatase, and glucose-6-phosphatase.
   a. **Pyruvate carboxylase:** Adds a CO_2 to pyruvate to create OAA (in the mitochondrial matrix)

KEY IDEA: This is the rate-limiting step of gluconeogenesis. In other words, gluconeogenesis proceeds only as fast or as slow as pyruvate carboxylase catalyzes this reaction.

- What increases flux through gluconeogenesis by stimulating pyruvate carboxylase?
  (1) Direct **acetyl CoA**
  (2) Indirect (all act by producing acetyl CoA)
      (a) **FFA** (can undergo beta oxidation to yield acetyl CoA)
      (b) **NADH** (inhibits citrate synthase, and thus creates ↑ levels of acetyl CoA)
  b. **PEPCK (phosphoenol-pyruvate carboxykinase):** Converts OAA to PEP (in the mitochondrial matrix) and releases 1 CO_2.
  c. **Fructose-1,6-bisphosphatase:** Converts fructose-1,6-BP to fructose-6-P (in the cytosol)
     - What increases flux through gluconeogenesis by stimulating F16-bisphosphatase?
       (1) Direct. **ATP**
       (2) Indirect
           (a) **Glucagon** (reduces inhibition of F16 bisphosphatase by ↓ production of F26BP)

(b) **Epinephrine** (same mechanism as glucagon)
(c) **Cortisol** (same mechanism as glucagon)
(d) **Caffeine** (same mechanism as glucagon)
- What <u>decreases</u> flux through gluconeogenesis by <u>inhibiting</u> F16 bisphosphatase?
  (1) <u>Direct</u>
    (a) **F26BP**
    (b) **AMP**
  (2) <u>Indirect</u>. **Insulin** (causes ↑ production of F26BP)
  d. **Glucose-6-phosphatase:** Converts glucose-6-P to glucose (in the cytosol)
6. **Clinical correlations**
   a. Why is it that <u>liver glycogen</u> yields glucose that can be used by the whole body, but <u>muscle glycogen</u> can only be used within muscle cells? → B/c the liver has glucose-6-phosphatase, and muscles do not. Therefore the liver can convert glucose-6-P (which cannot leave the cell) into glucose (which can).
   b. **Deficiency of liver glucose-6-phosphatase causes what disease?** → Von Gierke's disease (Type 1 glycogen storage disease). Have you already forgotten the beginning of the chapter?! ☺
7. **Summary of the regulation of gluconeogenesis**
   a. What are <u>all</u> the different substances that <u>increase</u> flux through gluconeogenesis?
     (1) <u>Direct</u>
       (a) **Acetyl CoA** (stimulates pyruvate carboxylase)
       (b) **ATP** (stimulates fructose-1,6-bisphosphatase)
     (2) <u>Indirect</u>
       (a) **FFAs.** How? → Inhibits pyruvate carboxylase (via creation of acetyl CoA from beta oxidation)
       (b) **NADH.** How? → Inhibits pyruvate carboxylase (via creation of <u>acetyl CoA</u> from the pathway: NADH → inhibition of citrate synthase → ↑ acetyl CoA (the first reaction of the TCA cycle)
       (c) **Glucagon.** How? → Three mechanisms that will be discussed later in this chapter
       (d) **Epinephrine.** How?
         - <u>Reduces inhibition of F16 bisphosphatase</u> (via ↓ production of F26BP)
         - <u>Inhibits the glycolytic enzyme pyruvate kinase</u> (which causes a buildup of PEP, which is then shunted into gluconeogenesis)
       (e) **Cortisol** (using the same two mechanisms as epinephrine)
       (f) **Caffeine** (using the same two mechanisms as epinephrine)
   b. What are <u>all</u> the different substances that <u>decrease</u> flux through gluconeogenesis?
     (1) <u>Direct</u>
       (a) **F26BP** (inhibits fructose-1,6-bisphosphatase)
       (b) **AMP** (inhibits fructose-1,6-bisphosphatase)

(2) Indirect
   (a) **Insulin**. How? → Discussed in the text that follows.
c. **Spotlight on insulin: What are the two ways that insulin decreases flux through gluconeogenesis?**
   (1) Decreases synthesis of four gluconeogenic enzymes (in the liver and kidneys)
      (a) Pyruvate carboxylase
      (b) PEPCK
      (c) Fructose-1,6-bisphosphatase
      (d) Glucose-6-phosphatase (in liver only)
   (2) Inhibits fructose-1,6-bisphosphatase (via ↑ production of F26BP)
d. **Spotlight on glucagon: What are the three ways that glucagon increases flux through gluconeogenesis?**
   (1) Increases synthesis of three gluconeogenic enzymes (in the liver and kidneys)
      (a) PEPCK
      (b) Fructose-1,6-bisphosphatase
      (c) Glucose-6-phosphatase
   (2) Reduces inhibition of F16-bisphosphatase (via ↓ production of F26BP)
   (3) Inhibits the glycolytic enzyme pyruvate kinase (which causes a buildup of PEP; PEP is then shunted into gluconeogenesis)

G. **Summary: The effects of insulin and glucagon on both glycolysis and gluconeogenesis**

> **Key idea:** There is a "switch" that profoundly affects **both** glycolysis and gluconeogenesis. This switch is called fructose-2,6-bisphosphate (F26BP).
> - When the switch is "on" (i.e., high levels of F26BP), the result is ↑ flux through glycolysis and ↓ flux through gluconeogenesis. When the switch is "off" (i.e., low levels of F26BP), then the reverse is true; ↓ flux through glycolysis and ↑ flux through gluconeogenesis.
> - What turns the switch "on?" → Insulin
> - What turns the switch "off?" → Glucagon, epinephrine, cortisol, caffeine

1. **Insulin**

> **Key idea:** Insulin stimulates glycolysis **and** inhibits gluconeogenesis.

a. **Insulin stimulates glycolysis. What are the three ways that it does this?**
   (1) Increases synthesis of three glycolytic enzymes in the liver:
      (a) Glucokinase
      (b) PFK-1
      (c) Pyruvate kinase
   (2) Stimulates PFK-1 (via ↑ production of F26BP)
   (3) Reduces inhibition of pyruvate kinase in the liver (via ↓ production of protein kinase A)

b. **Insulin inhibits gluconeogenesis. What are the two ways that it does this?**
   (1) Decreases synthesis of four gluconeogenic enzymes (in the liver and kidneys)
      (a) Pyruvate carboxylase
      (b) PEPCK
      (c) Fructose-1,6-bisphosphatase
      (d) Glucose-6-phosphatase (in liver only)
   (2) Inhibits fructose-1,6-bisphosphatase (via ↑ production of F26BP)
2. **Glucagon**

> **KEY IDEA: Glucagon inhibits glycolysis and stimulates gluconeogenesis.**

   a. **Glucagon inhibits glycolysis. What are the three ways that it does this?**
      (1) Decreases synthesis of three glycolytic enzymes in the liver
         (a) Glucokinase
         (b) PFK-1
         (c) Pyruvate kinase
      (2) Reduces stimulation of PFK-1 (via ↓ production of F26BP)
      (3) Inhibits pyruvate kinase in the liver (via ↑ production of protein kinase A)
   b. **Glucagon stimulates gluconeogenesis. What are the three ways that it does this?**
      (1) Increases synthesis of three gluconeogenic enzymes (in the liver and kidneys)
         (a) PEPCK
         (b) Fructose-1,6-bisphosphatase
         (c) Glucose-6-phosphatase
      (2) Reduces inhibition of F16 bisphosphatase (via ↓ production of F26BP)
      (3) Inhibits the glycolytic enzyme pyruvate kinase (which causes a buildup of PEP, which is then shunted into gluconeogenesis)

H. **SUMMARY: KEY SUBSTANCES AND RELATIONSHIPS IN CARBOHYDRATE METABOLISM**

> **KEY IDEA: If you do not get enough carbohydrates in your diet, what are the metabolic effects?**
> 1. Glycogen stores are depleted
> 2. ↑ Degradation of muscle proteins for use as energy
> 3. ↑ Breakdown of TG's; ↓ TG synthesis
> 4. ↑ Beta oxidation of FFAs; ↓ de novo synthesis of FFAs
> 5. ↑ Ketone production (in order to produce energy for the CNS)

1. **How do different parts of the body share energy with each other during periods of intense work?**

    a. **Gluconeogenesis in liver and kidneys.** Liver and kidneys help hard-working peripheral tissues (especially skeletal muscle) by doing gluconeogenesis and sending the resulting glucose to them.
    b. **Transfer of lactate from skeletal muscles to the liver and heart.** Skeletal muscles help the liver and heart by sending lactate to them (which is used in the TCA cycle after being first converted to pyruvate and then to acetyl CoA).
2. **Citrate**
   The three metabolic roles of citrate:
   a. **Source of acetyl CoA for de novo FFA synthesis** (via use of the citrate shuttle)
   b. **Stimulates de novo FFA synthesis** by stimulating acetyl CoA carboxylase in the cytoplasm
   c. **Shuts down glycolysis** by inhibiting PFK
3. **Pyruvate**
   The three sources of pyruvate:
   a. Glycolysis
   b. Deamination of certain amino acids
   c. Conversion from lactate
       (1) **In all cells besides the liver and kidneys, where does pyruvate go? (three fates)**
           (i) **Under low O_2 conditions—converts to lactate** to regenerate NAD^+
   - What enzyme converts pyruvate to lactate? → Lactate dehydrogenase
   - Lactate can then be sent to the liver for gluconeogenesis.
           (ii) Under **high O_2 conditions—enters mitochondrial matrix and gets decarboxylated** for entry into the Krebs cycle
   - What enzyme is responsible for pyruvate decarboxylation? → Pyruvate dehydrogenase
           (iii) **Converts to alanine** as a means of getting rid of intracellular NH_3
   - What enzyme converts pyruvate to alanine? → Aminotransferase
   - Alanine is then exported and sent to the liver where it either enters the urea cycle **or** is used for gluconeogenesis.
       (2) **In a liver or kidney cell, what happens to pyruvate? (four fates)**
   (Note: The first two fates are the same as in all cells; the last two are unique.)
           (i) Under **low O_2 conditions—converts to lactate** to regenerate NAD^+
   - What enzyme converts pyruvate to lactate? → Lactate dehydrogenase
           (ii) Under **high O_2 conditions—enters mitochondrial matrix and gets decarboxylated** to yield Acetyl Co-A for entry into the Krebs cycle
   - What enzyme is responsible for pyruvate decarboxylation? → Pyruvate dehydrogenase

(iii) **Converts to OAA to stimulate flux through the Krebs cycle** (liver and kidney only)
   - What enzyme converts pyruvate to OAA? → Pyruvate decarboxylase
   (iv) **Converts to OAA to enter gluconeogenesis** (liver and kidney only)
4. **Lactate**
   a. **Where does lactate originate?** → From the last step of anaerobic glycolysis—conversion of pyruvate to lactate by lactate dehydrogenase
   b. **Why is the creation of lactate under low O_2 conditions crucial for cell survival?**
      (1) **Short answer:** It saves the life of the cell by regenerating NAD^+ so that glycolysis can continue!
      (2) **Explanation:** Under low O_2 conditions, the ETC does not function at full capacity (because O_2 is required as the terminal electron acceptor). If ETC function is impaired, then reactions like the TCA cycle and beta oxidation are NOT helpful ways to generate energy (because the main way these reactions generate energy is by creating tons of NADH). Without the TCA cycle and beta oxidation, the cell relies exclusively on glycolysis to generate energy. Consequently, regeneration of cytoplasmic NAD^+ so that glycolysis can continue becomes absolutely crucial to cell survival!
   c. **What are the three fates of lactic acid in an exercising muscle?**
      (1) **Is built up** within the cell
      (2) Is sent to the liver or kidneys to be a substrate for **gluconeogenesis** (this is the first part of the "Cori cycle")
      (3) **Is sent to the liver and heart muscle** for entry into the TCA cycle (after first being reconverted to pyruvate)
   d. **Why is the buildup of lactic acid a bad thing?**
      (1) It causes a burning sensation in exercising muscles.
      (2) It is a weak acid and can thus lower blood pH (a condition called "lactic acidosis").
   e. **Clinical correlation**
      (1) **In what types of patients is lactic acidosis often seen?** → Those with poor perfusion of tissues (e.g., from a myocardial infarction, pulmonary embolism, or hemorrhagic shock).
      (2) Lactic acidosis results from the following pathway: Poor circulation → ↓ oxygen delivery to tissues → ↑ anaerobic glycolysis → ↑ production of lactate
5. **Acetyl CoA**
   a. **Repeat after me: "acetyl CoA cannot be used in gluconeogenesis, thus fat cannot be converted into glucose."** Good. Now, say it again three times... Trust me, this is important.
   b. **Where does acetyl CoA originate? (four sources)**
      (1) **Conversion from pyruvate** (by pyruvate dehydrogenase complex)
      (2) **Beta oxidation** of FFAs
      (3) **Catabolism of ketone bodies**
      (4) Deamination of the amino acid **isoleucine**

c. **Where does acetyl CoA go? (three fates)**
   (1) **Enters Krebs cycle** to generate energy (all cells)
   (2) Substrate for **de novo fatty acid synthesis** (only in liver, adipose, lactating mammary, and kidney cells)
   (3) Substrate for **ketone body synthesis** (which occurs only in liver cells)
d. **What metabolic effects are produced when acetyl CoA levels are high?**
   (1) ↑ **Flux through gluconeogenesis.** Acetyl CoA stimulates the first step of gluconeogenesis (conversion of pyruvate to OAA by pyruvate carboxylase)
   (2) ↑ **Flux through pathways that use acetyl CoA as a reactant**
      (a) Directly regulated pathways (i.e., acetyl CoA directly affects these reactions)
         • ↑ Flux through **TCA cycle**, thus ↑ energy generation (all cells)
         • ↑ **De novo synthesis of FFA** (liver, adipose, lactating mammary, and kidney cells)
         • ↑ **Ketone synthesis** (liver cells)
      (b) Indirectly regulated pathways (i.e., acetyl CoA affects these reactions via an intermediary molecule)
         • ↑ **TG synthesis** (liver, adipose, and brush border cells)
         • ↑ The pathway is [acetyl CoA] → ↑ [FFA] → ↑ TG synthesis
   (3) ↓ **Flux through pathways that produce acetyl CoA as a product**
      (a) Directly regulated pathways (i.e., acetyl CoA itself affects these reactions)
         • ↓ **Conversion of pyruvate to acetyl CoA** (all cells)
         • ↓ **Beta oxidation of FFA** (all cells)
      (b) Indirectly regulated pathways (i.e., acetyl CoA affects these reactions via an intermediary molecule)
         • ↓ Flux through **glycolysis** (all cells). (How? ↑ [Acetyl CoA] → ↑ **citrate** → ↑ inhibition of PFK → ↓ glycolysis)
         • ↓ **TG breakdown** (all cells). (How? ↑ [Acetyl CoA] → ↑ [FFA] → ↓ TG breakdown)
         • ↓ **Deamination of certain amino acids.** (How? ↑ [Acetyl CoA] → ↑ [pyruvate] → ↓ deamination of amino acids that yield pyruvate or acetyl CoA as their carbon skeleton)
e. **What metabolic effects are produced when acetyl CoA levels are low?** → The reverse of the previous listed reactions
   (1) ↓ **Flux through gluconeogenesis.** ↓ Acetyl CoA means ↓ stimulation of the first step of gluconeogenesis (conversion of pyruvate to OAA by pyruvate carboxylase)
   (2) ↓ **Flux through pathways that use acetyl CoA as a reactant**
      (a) Directly regulated pathways (i.e., acetyl CoA itself affects these reactions)
         • ↓ Flux through **TCA cycle**; therefore, ↓ energy generation (all cells)
         • ↓ **De novo synthesis of FFA** (liver, adipose, lactating mammary, and kidney cells)
         • ↓ **Ketone synthesis** (liver cells)

(b) Indirectly regulated pathways (i.e., acetyl CoA affects these reactions via an intermediary molecule)
      - ↓ **TG synthesis** (liver, adipose, and brush border cells). (How? ↓ [Acetyl CoA] → ↓ [FFA] → ↓ TG synthesis)
   (3) ↑ **Flux through pathways that produce acetyl-CoA as a product**
      (a) Directly regulated pathways (i.e., acetyl CoA directly affects these reactions)
         - ↑ **Conversion of pyruvate to acetyl CoA** (all cells)
         - ↑ **Beta oxidation of FFA** (all cells)
      (b) Indirectly regulated pathways (i.e., acetyl CoA affects these reactions via an intermediary molecule)
         - ↑ Flux through **glycolysis** (all cells). (How? ↓ [Acetyl CoA} → ↓ citrate → ↓ inhibition of PFK → ↑ glycolysis)
         - ↑ **TG breakdown** (liver, adipose, and brush border cells). (How? ↓ [Acetyl CoA] → ↓ [FFA] → ↑ TG breakdown)
         - ↑ **Deamination of certain amino acids.** (How? ↓ [Acetyl CoA] → ↓ [pyruvate] → ↑ deamination of amino acids that yield pyruvate or acetyl CoA as their carbon skeleton)
6. **The Cori cycle: A way for the liver to supply muscles with fresh glucose during periods of strenuous exercise.** The Cori cycle should be called the "glucose/lactate cycle".

KEY IDEA: Muscle cells export lactate and get glucose in return.
1. **In the muscle**
   (a) Muscle converts pyruvate to lactate (enzyme = lactate dehydrogenase).
   (b) Muscle sends lactate to liver.
2. **In the liver**
   (a) Liver sends lactate through gluconeogenesis.
   (b) Liver sends fresh glucose back to the muscle.

I. KETONE METABOLISM

1. **General principles**
   a. **What are the three ketone bodies?**
      (1) Acetoacetate
      (2) Beta-hydroxybutyrate
      (3) Acetone (cannot be metabolized for energy)
         - **Note:** Only two of the three ketone bodies can be metabolized for energy. Acetone cannot be.
   b. **Why are ketones helpful?**
      → As a supplemental energy source under low-glucose conditions
         - The basic idea is that ketones provide a way for the liver to <u>supply peripheral tissues with energy.</u>

KEY IDEA: Ketones are especially critical for the brain because the CNS has only two fuel sources: Glucose and ketones.

   c. **How do ketones provide energy to tissues?** → When <u>each ketone body</u> is degraded it <u>yields two acetyl CoA molecules</u>, which can then enter the TCA cycle.

d. Which two tissues actually prefer ketones to glucose?
   (1) Heart muscle
   (2) Renal cortex
2. **Ketone synthesis and degradation**

> **KEY IDEA:** What is the <u>only site</u> of ketone synthesis in the entire body? → The mitochondrial matrix of the liver

   a. **What is the rate-limiting enzyme of ketone synthesis?** → HMG CoA synthase
   b. **What is the basic idea of ketone synthesis?** → Combine two molecules of acetyl CoA
   c. **What is the basic idea of ketone degradation?** → Degrade the ketone body into two molecules of acetyl CoA (each of which can then enter the TCA cycle)
   d. <u>When</u> **does the liver produce ketones?** → The liver is <u>always</u> producing <u>low levels</u> of ketone bodies.
      • Why? → B/c ketone production occurs whenever the acetyl CoA concentration exceeds the liver's capacity to shuttle acetyl CoA into the TCA cycle.

> **KEY IDEA:** The liver produces ketone bodies, but it, itself, cannot use them for energy.
> • Why not? → B/c the liver <u>lacks thiophorase</u> (an enzyme which catalyzes the first step of ketone degradation).

   e. **At the molecular level, what two things stimulate ketone synthesis?**
      (1) Accumulation of **acetyl CoA** (this is the main stimulus)
      (2) **Epinephrine or norepinephrine** from the adrenal medulla

> **KEY IDEA:** Remember that the <u>main determinant</u> for ketone production is intracellular levels of <u>acetyl CoA</u>.
> • If intracellular (acetyl CoA) is <u>high</u>, ketone synthesis is great.
> • If intracellular (acetyl CoA) is <u>low</u>, ketone synthesis is minimal.

3. **Clinical scenarios involving ketone metabolism**
   a. **Clinical correlation** → What are the two prototypical clinical scenarios in which you see ketone levels that are <u>significantly increased</u>?
      (1) Sustained hypoglycemia
      (2) Untreated type 1 diabetes
   b. **Clinical correlation** → **How does hypoglycemia (e.g., starvation, a very low-carbohydrate diet) cause ↑ ketone production?** → By causing <u>accumulation of acetyl CoA</u> via the following three pathways:
      (1) ↓ Carbohydrates → ↓ flux through glycolysis → ↓ pyruvate → ↓ conversion of pyruvate to OAA → ↓ OAA → accumulation of acetyl CoA

(2) ↓ Carbohydrates → ↑ gluconeogenesis → ↑ use of OAA in gluconeogenesis → ↓ OAA → accumulation of acetyl CoA
(3) ↓ Carbohydrates → ↑ beta oxidation of FFAs → accumulation of acetyl CoA
c. When the (ketone) in the <u>blood</u> rises, what is this called? → "Ketonemia"
d. When the (ketone) in the <u>urine</u> rises, what is this called? → "Ketonuria"

> **KEY IDEA:** Why is a buildup of ketone bodies inconvenient or harmful or both?
> 1. Inconvenient—causes bad breath
> 2. Harmful—can cause ketoacidosis
>   - What is "ketoacidosis?" → A state in which the blood becomes acidic because of an <u>accumulation of ketone bodies</u> (which are weak acids) combined with <u>decreased plasma volume</u>. Explanation: Ketones are weak acids. Additionally, ketones excreted in the urine act as an osmole and draw water with them, causing a ↓ in plasma volume. Therefore ↑ amounts of H^+ combined with ↓ plasma volume can create a severe acidosis called "ketoacidosis."

J. **DIABETES: TOO MUCH GLUCOSE IN THE BLOOD**
   1. Key principles about <u>all</u> types of diabetes (<u>b</u>oth type 1 and type 2)
      a. **What is the leading cause of adult blindness and amputation?**
         → Diabetes
         - It is also a major cause of renal failure, heart attacks, and stroke.
      b. **What causes the excessive thirst (polydipsia) and excessive urination (polyuria) in patients with diabetes?** ↑ Glucose in blood → ↑ blood and urine osmolarity → ↑ entry of water into blood vessels and urine → ↑ thirst and ↑ urination
      c. **What causes cataracts and peripheral neuropathy in a patient with diabetes?**
         (1) **Short answer:** → Intracellular accumulation of sorbitol, which causes the cell to swell with water.
         (2) **Explanation:** Intracellular glucose is converted to sorbitol by aldose reductase. Because sorbitol cannot leave the cell, it acts as an osmole that attracts water—thus causing swelling. This swelling, when it occurs in the lens, is thought to cause cataracts; when it occurs in the Schwann cells of peripheral nerves, is thought to cause neuropathy.
   2. **A comparison of the two types of diabetes mellitus** (Table 10.2)

TABLE 10.2

	Insulin-Dependent Diabetes Mellitus (IDDM)	Non–Insulin-Dependent Diabetes Mellitus (NIDDM)
Synonym	Type 1—Juvenile onset	Type 2—Adult onset
Prevalence	Rare (10%-20% of diabetics)	Most common (80%-90% of diabetics)

Continued

Table 10.2—cont'd

	Insulin-Dependent Diabetes Mellitus (IDDM)	Non–Insulin-Dependent Diabetes Mellitus (NIDDM)
Basic problem	Lack of insulin production	(1) Cells become insensitive to insulin (2) Pancreatic beta cells cannot secrete enough insulin to produce an effect
Cause	Autoimmune destruction of pancreas beta cells (triggered by a viral infection or environmental factors)	Not precisely known. However, obesity plays some role in producing insulin insensitivity
Genetic predisposition?	Moderate	VERY STRONG!
Age of onset	Childhood or puberty	Usually after age 35
Classic symptoms	• Frequent urination (polyuria) • Excessive thirst (polydipsia) • Excessive hunger (polyphagia) • Patient is malnourished	• Frequent urination (polyuria) • Excessive thirst (polydipsia) • Patient is significantly overweight Note: Excessive hunger (polyphagia) is less frequently seen than in Type 1 diabetes.
Biochemical signs	• Hyperglycemia (fasting glucose > 140 mg/dL) • Ketoacidosis	• Hyperglycemia (fasting glucose > 140 mg/dl)
Plasma (insulin)	Low or absent (B/c beta cells have been destroyed)	Normal or high (B/c beta cells are functional)
If left untreated, what acute complication are you worried about?	• Ketoacidosis causing coma or death	• Hyperosmolar coma
Chronic complications	• Atherosclerosis • Retinopathy • Nephropathy • Neuropathy • Hypertriglyceridemia • Impaired ability to secrete glucagon and epinephrine in response to low blood sugar	• Atherosclerosis • Retinopathy • Nephropathy • Neuropathy
Ketoacidosis?	Common. Why? B/c ↓ insulin and ↑ glucagon stimulate ketone formation	Rare
Goal and strategies of treatment	• Keep blood glucose at normal levels through administration of insulin	• Lower blood glucose levels (often through weight reduction and/or sulfonylurea drugs)
Requires treatment with insulin?	Yes—for the remainder of life	Usually not

Table 10.2—cont'd

	Insulin-Dependent Diabetes Mellitus (IDDM)	Non–Insulin-Dependent Diabetes Mellitus (NIDDM)
Responsive to oral hypoglycemic drugs (e.g., sulfonylureas)?	NO	YES

KEY IDEAS: Type 1 diabetes

- **Important note:** The ↑ blood [glucagon] seen in a person with type 1 diabetes is transitory. Glucagon is elevated initially, but by 4 years after diagnosis, people with diabetes almost universally suffer from an inability to secrete glucagon.
- **Why do people with type 1 diabetes develop hyperglycemia?** → As a result of ↓ [insulin] and ↑ [glucagon]. Here is the main pathway: ↓ [insulin] and ↑ [glucagon] → ↑ gluconeogenesis, ↑ glycogenolysis, and ↓ glucose uptake into cells → ↑ blood glucose
- **Why are people with type 1 diabetes at risk for ketoacidosis?** → As a result of ↓ [insulin] and ↑ [glucagon]. Here is the main pathway: ↓ [insulin] and ↑ [glucagon] → ↑ TG breakdown and ↑ beta oxidation of FFAs → ↑ acetyl CoA → ↑ ketone formation
- **Why do people with type 1 diabetes develop hypertriglyceridemia?** → As a combination of ↓ [insulin], ↑ [glucagon], and ↓ lipoprotein lipase activity. Here are the two steps of the pathway:
  1. ↓ [Insulin] and ↑ [glucagon] → ↑ TG breakdown in the liver → ↑ FFA → ↑ Packaging of FFAs into VLDLs and chylomicrons (for delivery to peripheral tissues)
  2. ↑ Packaging of FFAs into VLDLs and chylomicrons + ↓ lipoprotein lipase activity = hypertriglyceridemia
- **How is type 1 diabetes similar to starvation? How is it different?**
  Similar
  - In both cases, **cells are not receiving enough glucose**; thus they "feel" and "act" as if they are starving.
  - There is an **increased breakdown of muscle proteins and fat** to use them for energy.
  - The liver **increases ketone production**.

  Different
  - **Insulin in a person with type 1 diabetes is virtually absent**; but is present during starvation in low amounts.
  - A person with **type 1 diabetes has hyperglycemia**; a starving person maintains blood glucose near normal levels.
  - **Type 1 diabetes involves a much greater ↑ in ketone formation**. Therefore the ketoacidosis than that of a person with type 1 diabetes is much more severe compared with a starving person.

Continued

KEY IDEAS: Type 1 diabetes—cont'd
Why should a person with type 1 diabetes avoid activation of the sympathetic nervous system? → B/c circulating epinephrine would inhibit the release of what little insulin they are capable of producing. Remember that epinephrine inhibits insulin release. This effect may be strong enough to completely clear their system of insulin, causing major problems for cells that depend on insulin for glucose uptake (e.g., all cells except the brain, liver, and RBCs).
- **Why are people with <u>chronic</u> diabetes at risk for <u>hypoglycemia</u> in response to treatment?**
  Short answer: B/c over time they lose the ability to secrete glucagon and epinephrine in response to treatment-induced hypoglycemia.
  Long answer: Treating diabetes can have the unintended side effect of creating a <u>hypoglycemic</u> state. Such hypoglycemia is normally offset by secretion of glucagon or epinephrine (both of which cause glucose entry into the blood). However, people with <u>chronic</u> type 1 diabetes lose the ability to secrete glucagon and epinephrine in response to hypoglycemia. So, essentially, the induction of a hypoglycemic state is dangerous for them because they lack the hormonal mechanisms to compensate.

KEY IDEAS: Type 2 diabetes
- **How is hyperglycemia produced in a person with type 2 diabetes?** → B/c of cellular resistance to insulin thus resistance to insulin-mediated glucose uptake. Here is the pathway: ↓ sensitivity to insulin → ↓ glucose uptake into cells → hyperglycemia
- **Why is weight reduction a treatment for hyperglycemia in people with type 2 diabetes?** → B/c <u>weight loss improves cellular sensitivity to insulin.</u> Here is the pathway: greater sensitivity to insulin → ↑ glucose uptake into cells → ↓ hyperglycemia

CHAPTER 11
LIPID AND CHOLESTEROL METABOLISM

A. GENERAL PRINCIPLES

1. **There are three types of lipids:**
   a. **Sterols:** Multi-ring structures with side chains attached
      ex) Cholesterol
   b. **Phospholipids:** A glycerol or sphingosine backbone with two fatty acids and a polar group
      ex) Phospholipid bilayer that forms cell membranes
   c. **Triglycerides:** A glycerol backbone with three fatty acids
      - TGs are 95% of the "fat" we consume in the diet.
      - TGs are 100% of stored body fat (i.e., adipose tissue).

2. **Important functions of lipids in our bodies:**
   a. Help with absorption of fat-soluble vitamins (vitamins A, D, E, and K)
   b. Precursors for prostaglandin, prostacyclin, leukotriene, and thromboxane synthesis
   c. Precursors for steroid hormone synthesis
   d. Supply energy (FFAs enter TCA cycle; glycerol-P enters glycolysis or gluconeogenesis)
   e. Provide insulation
   f. "Brown fat" produces heat
   g. Slow gastric emptying
   h. Provide a sense of "fullness" after a meal (by stimulating CCK release)
   i. Give food a desirable taste and texture

B. ALL ABOUT FATTY ACIDS

1. **Fatty acids (FAs) travel in the blood bound to serum albumin**
2. **Esterified versus unesterified ("free") fatty acids**
   a. Unesterified = soluble in water.
   b. Esterified = INsoluble in water (The same is true for cholesterol.)

KEY IDEA: The body esterifies things to make them LESS soluble.

3. **Why are fatty acids such a helpful way to store energy?**
   a. Can be concentrated in small spaces
   b. Can enter TCA cycle
   c. Can produce glucose, if needed (glycerol can enter gluconeogenesis)
   d. Can produce ketone bodies, if needed (beta oxidation of FFA → acetyl CoA → ketones)
   e. **Which three tissues cannot use plasma FFAs to generate energy?**
      (1) The brain
      (2) RBCs
      (3) Adrenal medulla
4. **FAs classified by structure—saturated, monounsaturated, or polyunsaturated**
   a. Saturated = no double bonds
   b. <u>Mono</u>unsaturated = one double bond
   c. <u>Poly</u>unsaturated = more than one double bond
5. **What are the essential fatty acids?** → "Essential" = cannot be synthesized by the body, thus they must be present in the diet. **There are two essential fatty acids (both of which are also polyunsaturated):**
   a. **Lino<u>leic</u> acid** (an omega 6)
      (1) A precursor for prostaglandin synthesis
      (2) A good dietary source = seeds, nuts
   b. **Lino<u>lenic</u> acid** (an omega 3). A good dietary source = fish oil

C. **DE NOVO FFA SYNTHESIS: AN ANABOLIC PATHWAY THAT TAKES ACETYL CoA AND TURNS IT INTO PALMITATE**

> <u>KEY IDEA</u>: **De novo synthesis of FFAs from acetyl CoA occurs in the cytosol of liver, lactating mammary gland, adipose, and kidney cells. The most important enzymes involved are acetyl CoA carboxylase and fatty acid synthase. The process requires NADPH.**

1. **Metabolic context:** De novo FFA synthesis occurs when the body has plenty of energy and wants to synthesize FFA for storage or transport to other tissues.
2. **Reaction summary:** acetyl CoA + NADPH → palmitate → all other FFAs
   a. **Starting reactants and where they come from:**
      (1) Acetyl CoA (comes from various sources)
      (2) NADPH (comes from HMP shunt or from malate dehydrogenase reaction)
   b. **End-products and where they go:**
      • Palmitate or other FFA (Its fate is to get broken down into acetyl CoA via beta oxidation or to get used to synthesize a TG.)
3. **Basic strategy of the pathway:** Convert some acetyl CoA molecules into malonyl CoA. Then link acetyl CoA and malonyl CoA molecules together to form a growing chain

- How many NADPHs are used per each addition of acetyl CoA into the growing fatty acid chain? → 2 NADPH per acetyl CoA
4. Two key enzymes involved:
   a. **Acetyl CoA carboxylase:** Turns acetyl CoA into malonyl CoA
      (acetyl CoA + CO_2 + ATP → malonyl CoA)

KEY IDEA: This is the **rate-limiting step and the only major control point in de novo FFA synthesis!**

- What cofactor does acetyl CoA carboxylase require to work? → Biotin
   b. **Fatty acid synthase complex:** Links acetyl CoA and malonyl CoA molecules together to form a growing chain
      (Acetyl CoA + malonyl CoA → palmitate)
5. Stimulation and inhibition of de novo FFA synthesis
   a. These two substances ↑ rate of de novo fatty acid synthesis
      (1) **Citrate** (by stimulating acetyl CoA carboxylase)
      (2) **Insulin** (dephosphorylation of acetyl CoA carboxylase thus activating it)
   b. These four substances ↓ the rate of de novo fatty acid synthesis
      (1) **Malonyl CoA** (by inhibiting acetyl CoA carboxylase)
      (2) **Palmitoyl CoA** (by inhibiting acetyl CoA carboxylase)
      (3) **Glucagon** (phosphorylation of acetyl CoA carboxylase thus inactivating it)
      (4) **Epinephrine** (phosphorylation of acetyl CoA carboxylase thus inactivating it)
   c. Think about it logically. End-products: Malonyl and Palmitoyl inhibit de novo synthesis by feedback inhibition of the loop. When you are stressed studying for boards or exercising, epinephrine kicks in, and your body has no time to be making fatty acid. And glucagon is released under conditions of low blood sugar. If you have low blood sugar, it is not time to synthesize FAs; it is time to degrade them.
6. Problem: The <u>enzymes</u> for de novo fatty acid synthesis are all in the <u>cytoplasm</u>, but the <u>starting material</u> (acetyl CoA) is in the mitochondrial <u>matrix</u> and cannot escape. What to do?
   - Solution: The citrate shuttle.
      (1) Step 1: Acetyl CoA is converted to citrate (by citrate synthase).
      (2) Step 2: Citrate leaves the mitochondria and enters the cytoplasm.
      (3) Step 3: Citrate is reconverted into acetyl CoA (by citrate lyase) (Figure 11.1).
7. **One of the effects of ↑ flux through the de novo FFA synthesis pathway is to inhibit beta oxidation. How does this happen?**
   → The malonyl CoA produced by de novo synthesis inhibits the carnitine shuttle (which shuttles FFA into the mitochondria for beta oxidation).

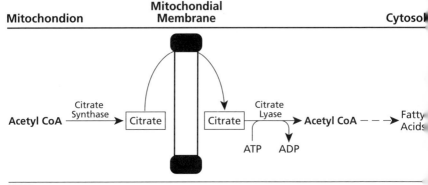

FIGURE 11.1 The citrate shuttle transports acetyl CoA into the cytoplasm during de novo FFA synthesis.

- **How does malonyl CoA inhibit the carnitine shuttle?** → By inhibiting carnitine acyltransferase 1
8. **Nutritional correlations:**
   a. Carbohydrate consumption in the diet stimulates fat synthesis and storage by increasing the levels of both acetyl CoA carboxylase and fatty acid synthase. Observe the following pathway:
      - Prolonged consumption of a high-carbohydrate or fat-free diet → ↑ levels of these enzymes → ↑ de novo fatty acid synthesis → ↑ TG synthesis and storage
   b. The previous fact is a rationale used to support the low-carbohydrate theory of dieting, in which one would lose fat by reversing that pathway: ↓ carbohydrate consumption → ↓ levels of these enzymes → ↓ fatty acid synthesis → ↓ TG synthesis and storage → attractive physique

D. BETA OXIDATION: A CATABOLIC PATHWAY THAT TAKES AN FFA AND REPEATEDLY REMOVES TWO CARBON FRAGMENTS TO MAKE ACETYL CoA, NADH, AND FADH$_2$

> **KEY IDEA:** Beta oxidation of FFAs into acetyl CoA occurs in the mitochondrial matrix of all cells except RBCs and the brain. The most important enzymes involved are carnitine acyltransferase 1 and 2. The process creates lots of NADH and FADH$_2$.

1. **Metabolic context:** Beta oxidation occurs when the body is low on energy. Conversion of FFA into acetyl CoA essentially allows entry of fats into the TCA cycle.
2. **Basic strategy of the pathway:** Repeatedly remove 2-carbon fragments from an FFA chain. Convert these 2-carbon fragments into acetyl CoA. If you get to the end of the chain and there are three carbons left, then you convert this 3-carbon fragment (propionyl CoA) into succinyl CoA. **At the end of each round of beta oxidation, what three things are released?**

a. 1 Acetyl CoA
   b. 1 FADH
   c. 1 NADH
3. **Reaction summary:** FFA → 8 acetyl CoA + 7 NADH + 7 FADH$_2$ **(or)** FFA → 7 acetyl CoA + 1 succinyl CoA + 7 NADH + 7 FADH$_2$
   a. **Starting reactants and where they come from:** Fatty acyl CoA (comes from the breakdown of a TG).
   b. **End-products and where they go:**
      (1) If the fatty acid has an <u>even number</u> of carbons, the end-product is **acetyl CoA** (which has three different fates).
      (2) If the fatty acid has an <u>odd number</u> of carbons, the end-product is **succinyl CoA** (which enters the TCA cycle).
4. **Keeping track of the indirect ATP production (via creation of NADH and FADH$_2$)**
   a. **How many ATP are generated per acetyl CoA produced by beta oxidation?** → 5
   b. **How many ATP are generated per acetyl CoA that gets produced <u>and</u> goes through the TCA cycle?** → 12
5. **Four key enzymes are involved in beta oxidation:**
   a. Carnitine acyltransferase 1 **and** 2 (involved in the carnitine shuttle)

KEY IDEA: Carnitine acyltransferase 1 = rate-limiting enzyme

   b. Acyl CoA dehydrogenase (generates FADH$_2$)
   c. Beta-hydroxyacyl CoA dehydrogenase (generates NADH)
6. **Simulation and inhibition of beta oxidation**
   a. **What stimulates beta oxidation?** → Glucagon
   b. **What inhibits beta oxidation?**
      (1) Direct inhibitor: **Malonyl CoA**
      (2) Indirect inhibitor: Flux through de novo FFA synthesis (which produces malonyl CoA)
         - **How does malonyl CoA inhibit beta oxidation?** → By inhibiting the carnitine shuttle (via inhibition of carnitine acyltransferase 1)
         - **Why does it make sense that malonyl CoA should inhibit beta oxidation?**
            → B/c malonyl CoA is an important substrate for de novo FFA synthesis. High [malonyl CoA] indicates increased flux through the de novo fatty acid synthesis pathway. Why break down fatty acids (via beta oxidation) at the same time you are creating them (via de novo fatty acid synthesis)?
7. **Problem: The <u>enzymes</u> for beta oxidation are all in the <u>mitochondrial matrix</u>, but the <u>starting material</u> (fatty acyl CoA) is in the <u>cytoplasm</u> and cannot enter. What to do?**
   → **Solution: The carnitine shuttle.**
   a. Step 1: Carnitine reacts with cytoplasmic fatty acyl CoA to produce O-acylcarnitine (enzyme = carnitine acyl transferase 1).
   b. Step 2: O-acylcarnitine leaves the cytoplasm and enters the mitochondrial matrix.

c. Step 3: O-acylcarnitine reacts to reform both fatty acyl CoA <u>and</u> carnitine (enzyme = carnitine acyl transferase 2).
d. Step 4: Carnitine returns to the cytoplasm (Figure 11.2).

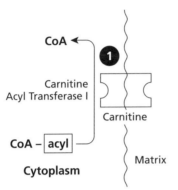

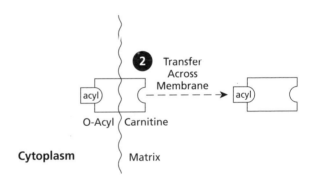

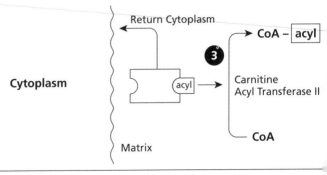

FIGURE 11.2 Carnitine shuttle.

8. **Summary: How do fatty acids produce energy by entering the TCA cycle? (two possible ways)**
   a. FAs with an <u>even number of carbons</u> get converted to acetyl CoA, which enters the TCA cycle at the beginning.
   b. FAs with an <u>odd number of carbons</u> get broken down to a 3-carbon structure, **succinyl CoA,** which then enters the TCA cycle in the middle.

E. SUMMARY AND COMPARISON OF DE NOVO FFA SYNTHESIS VERSUS BETA OXIDATION (Table 11.1)

TABLE 11.1 De Novo FFA Sythesis versus Beta Oxidation

	De Novo FFA Synthesis	Beta Oxidation of FFA
Summary of pathway	Acetyl CoA → FFA	FFA → Acetyl CoA
Purpose	Use surplus energy substrate to synthesize fat	Convert fat into a form that allows it to generate energy via the TCA cycle
Period of greatest flux	After a meal	Starvation
Location	Cytoplasm of liver, lactating mammary, adipose, and kidney cells	Mitochondrial matrix of all cells except brain and RBCs
Shuttle system used	Citrate shuttle (used to carry acetyl CoA from the mitochondria into the cytosol)	Carnitine shuttle (used to carry FFAs from the cytoplasm into the mitochondria)
Main enzymes (rate limiting enzyme in bold)	• **Acetyl CoA carboxylase** • Fatty acid synthase complex	• **Carnitine acyltransferase 1 and 2** • Acyl CoA dehydrogenase (generates $FADH_2$) • Beta-hydroxyacyl CoA dehydrogenase (generates NADH)
Cofactor	NADPH	NAD+ and FAD
Stimulated by	• Citrate (stimulates acetyl-CoA carboxylase) • Insulin (stimulates creation of active form of acetyl CoA carboxylase)	Glucagon
Inhibited by	• Intracellular (FFA) (e.g., malonyl CoA, palmitoyl CoA) (inhibits acetyl CoA carboxylase) • Glucagon (stimulates inactivation of acetyl CoA carboxylase) • Epinephrine or norepinephrine (stimulates inactivation of acetyl CoA carboxylase)	DIRECT INHIBITOR • Malonyl CoA (inhibits carnitine acyltransferase 1) INDIRECT INHIBITOR (via production of malonyl CoA) • Flux through de novo FFA synthesis pathway

F. Key principles of TG storage and metabolism

TG synthesis and breakdown occur simultaneously in cells.
1. What two enzymes catalyze TG synthesis? → Fatty acyl CoA synthetase and acyltransferase
2. What enzyme catalyzes TG breakdown? → Hormone sensitive lipase

> **Key idea:** Hormones determine whether net TG breakdown or synthesis occurs by stimulating or inhibiting hormone sensitive lipase.

G. TG synthesis ("lipogenesis"): An anabolic pathway that converts three FFAs and one glycerol phosphate into a triglyceride

> **Key idea:** Creation of a TG from three FFAs and one glycerol phosphate (i.e., TG synthesis) occurs in the cytoplasm of liver, adipose, and brush border cells. The enzymes involved are fatty acyl CoA synthetase and acyltransferase.

1. **Metabolic context:** You just ate a meal. Circulating insulin is stimulating TG synthesis.
2. **Summary of the reaction:** Three FFAs + one glycerol phosphate → one TG
3. **Basic strategy of the pathway:** First, attach coenzyme A to fatty acid to create fatty acyl CoA. Then, attach three fatty acyl CoAs to a glycerol-phosphate backbone.
4. **Key enzymes and the reactions they catalyze**
   a. **Step 1: Fatty acyl CoA synthetase** (a.k.a., thiokinase): Attaches a CoA group to each fatty acid

   $$\text{fatty acid} \xrightarrow{\text{fatty acyl CoA synthetase}} \text{fatty acyl CoA}$$

   b. **Step 2: Acyltransferase:** Attaches three fatty acyl CoAs to a glycerol phosphate backbone

   $$\text{three fatty acyl CoAs + one glycerol phosphate} \xrightarrow{\text{acyl transferase}} \text{one TG}$$

5. **Simulation and inhibition of TG synthesis**
   a. **What hormone ↑ the net rate of TG synthesis?** → Insulin. **How?** → Via dephosphorylation (and thus, inactivation) of hormone sensitive lipase
   b. **What five hormones ↓ the net rate of TG synthesis?**
      (1) **Glucagon.** How? → Via phosphorylation (and thus activation) of hormone sensitive lipase

(2) Cortisol
   (3) Epinephrine and norepinephrine
   (4) Thyroid hormones
   (5) HGH
  c. Think about this logically again. You are stressed studying for boards. Epinephrine, cortisol, and other hormones kick in. Your body has no time to be making TGs or fat. It needs all the energy it can get to go to the brain, heart, and skeletal muscles.

H. TG STORAGE: TAKES NEWLY SYNTHESIZED TGS AND STORES THEM IN ADIPOSE CELLS

1. **What is the location of almost all TG storage?** → Adipose cells
2. **Do muscles have stored TGs?** → No

KEY IDEA: Does the liver have stored TGs? → **Not really!** The liver is a major site of TG synthesis, but NOT storage. The liver stores only a small quantity of TGs. Most of the TGs synthesized by the liver are packaged into VLDLs for shipment to peripheral tissues.

I. TG DEGRADATION ("LIPOLYSIS"): BREAKING DOWN TGS TO YIELD THREE FFAS AND ONE GLYCEROL

KEY IDEA: Degradation of a TG into three FFAs and one glycerol occurs in two locations:
1. **On cell membrane surfaces (enzyme = lipoprotein lipase)**
2. **In the cytoplasm (enzyme = hormone sensitive lipase)**
   TG → 3 FFA + glycerol

1. **Metabolic context:** The body is low on energy and decides to use some stored TGs. How does it do this? The first step is to degrade the TG and release the three FFAs. After that, the three FFAs can enter beta oxidation for eventual conversion into acetyl CoA and entry into the TCA cycle.
2. **TG breakdown, location #1: Cell membrane surfaces**

KEY IDEA: Breakdown of TGs happens on the outer surface of all cell membranes, and is catalyzed by lipoprotein lipase. The glycerol and FFAs that get released either diffuse into nearby cells, or bind to serum albumin for transport to other tissues.

3. **TG breakdown, location #2: the cytoplasm**

KEY IDEA: Breakdown of TGs also happens in the cytoplasm of liver and adipose cells and is catalyzed by hormone sensitive lipase.

4. **Stimulation and inhibition of TG degradation**
   a. **What five hormones ↑ the rate of TG breakdown in liver and adipose cells by affecting hormone sensitive lipase?**

(1) **Glucagon.** How? → Phosphorylation (and thus, activation) of HSL
(2) **Epinephrine and norepinephrine.** How? → Phosphorylation (and thus activation) of HSL
(3) **Thyroid hormones.** How? → Indirectly, by enhancing sensitivity of cells to epinephrine or norepinephrine
(4) **Cortisol.** How? Phosphorylation (and thus activation) of HSL
(5) **HGH.** How? → Phosphorylation (and thus activation) of HSL
  b. Think about it logically. You are stressed for the third time. You want to release fat energy → epinephrine, cortisol, STRESS from boards kick in!!! → Activate my HSL!
  c. **What two substances ↓ rate of TG breakdown by inhibiting HSL?**
(1) **Insulin.** How? → Dephosphorylation (and thus inactivation) of HSL
(2) **Blood glucose.** How? → Indirectly, by stimulating insulin release

J. EXOGENOUS AND ENDOGENOUS TG TRANSPORT
1. **Exogenous versus endogenous TG synthesis**
  a. <u>Exo</u>genous = made <u>outside</u> of the body
  b. <u>Endo</u>genous = made <u>inside</u> the body (i.e., in liver, adipose, brush border cells)

KEY IDEA: Both exogenous and endogenous TG transport use lipoproteins (e.g., chylomicrons, VLDLs). Lipoproteins have the following structure:
- Outer membrane = phospholipids, proteins, some free cholesterol
- Interior core = TGs and cholesterol esters

2. **EXOGENOUS TG transport: Let's say you pig out and get some TGs from the diet. How do they get to the tissues that need them?**
  a. TGs get **packaged into chylomicrons** and secreted into the lymph system. (Which apoprotein mediates chylomicron secretion? → **B-48**.) Chylomicrons flow through the lymph system and join the bloodstream in the superior vena cava.
  b. Once they join the bloodstream, **chylomicrons are modified by HDL.** (What does HDL do to them? → Adds apoproteins C-2 and E. Why does this make sense? → B/c the chylomicron will need C-2 to bind to lipoprotein lipase and E for the eventual remnants to be taken up by the liver.)
  c. After modification, chylomicrons travel to peripheral tissues. Those **tissues that need TGs bind and hydrolyze chylomicrons** using the cell surface enzyme lipoprotein lipase. (Which apoprotein activates lipoprotein lipase? → **C-2**.) As the chylomicron is degraded by lipoprotein lipase, it releases TGs, which diffuse into the cell.
  d. **Once the chylomicron releases its TGs at the cell surface, what two things happen?**
(1) Any **free cholesterol** that spills into the blood is **esterified by plasma LCAT and then scooped up by HDL** and brought back to the liver.
(2) **The chylomicron undergoes "remodeling" by HDL**

e. HDL basically does four things to it:
　　　(1) **HDL <u>removes</u> part of the cell membrane.** Why does this make sense? → B/c once the chylomicron releases all of its TGs, it is left with a big, floppy membrane covering a lot of empty space inside. (Kind of how a person who quickly loses a lot of weight is left with loose, floppy skin)
　　　(2) **HDL <u>removes</u> any remaining TGs.**
　　　(3) **HDL <u>removes</u> apoprotein C-2.** (Why does this make sense? → Since the chylomicron has already reacted with lipoprotein lipase, it no longer needs Apo C-2. Therefore HDL removes Apo C-2 and recycles it by giving it to someone who <u>will</u> need it—newly formed chylomicrons emerging from the lymph system or newly formed VLDL emerging from the liver.)
　　　(4) **HDL <u>adds</u> cholesterol esters into the middle of the chylomicron.** (Which apoprotein mediates transfer of cholesterol esters from HDL? → CTEP.)
　　f. **Once chylomicron remodeling by HDL is completed, what are the new structures called?** → Chylomicron remnants
　　g. **What is the fate of these chylomicron remnants?** → Travel to the liver and get absorbed by endocytosis. (Which two apoproteins mediate uptake of chylomicron remnants by the liver? → **B-48 and E**.) Once absorbed, chylomicron remnants are degraded by lysosomes and certain parts (e.g., cholesterol, phospholipids) are recycled.
　　h. **What are chylomicron remnants contributing to the liver as they get absorbed?** → Cholesterol. (Remember that chylomicron remnants are packed full of cholesterol esters, some of which originally came from the diet and some of which were added by HDL during remodeling.)
　　i. **Summary:** Dietary TGs → gut → chylomicrons → lymph → bloodstream → peripheral tissues → chylomicron remnants recirculate → liver
　　j. **Take-home message:** The transport of EXOGENOUS TGs <u>from</u> the GI tract <u>to</u> peripheral tissues uses <u>chylomicrons</u> as a carrier.
3. **ENDOGENOUS TG transport, SCENARIO #1: Your peripheral tissues are working hard and need TGs for energy. Because the liver is the major site of TG synthesis, how do the TGs get <u>from</u> the liver <u>to</u> the peripheral tissues?**
　　a. By being **packaged into VLDLs**, which are then secreted by liver into the bloodstream. (Which apoprotein mediates VLDL secretion? → **B-100**.)
　　b. After they join the bloodstream, **VLDLs are modified by HDL.** (What does HDL do to them? → Adds apoproteins C-2 and E. Why does this make sense? → B/c the VLDL will need C-2 to bind to lipoprotein lipase, and E for the eventual remnants to be taken up by the liver.)
　　c. VLDLs travel to peripheral tissues. Those **tissues that need TGs bind and hydrolyze VLDLs** using the cell surface enzyme lipoprotein lipase. (Which apoprotein activates lipoprotein lipase? → **C-2**.) As the VLDL is degraded by lipoprotein lipase, it releases TGs, which diffuse into the cell.

d. **After the VLDL releases its TGs at the cell surface, what two things happen?**
   (1) Any **free cholesterol** that spills into the blood **is esterified by plasma LCAT and scooped up by HDL** and brought back to the liver.
   (2) **The VLDL undergoes "remodeling" by HDL.** HDL basically does four things to it:
      (a) **HDL removes part of the cell membrane.** Why does this make sense? → B/c once the VLDL releases all of its TGs, it is left with a big, floppy membrane covering a lot of empty space inside. (Kind of how a person who quickly loses a lot of weight is left with loose, floppy skin.)
      (b) **HDL removes any remaining TGs.**
      (c) **HDL removes apoprotein C-2.** (Why does this make sense? → B/c the VLDL has already reacted with lipoprotein lipase, it no longer needs Apo C-2. Thus HDL removes Apo C-2 and recycles it by giving it to someone who will need it—newly formed chylomicrons emerging from the lymph system or newly formed VLDL emerging from the liver.)
      (d) **HDL adds cholesterol esters into the middle of the VLDL.** (Which apoprotein mediates transfer of cholesterol esters from HDL? → **CTEP.**)
e. **After remodeling by HDL is completed, what are the new structures called?** → IDLs (intermediate density lipoproteins). These are essentially "VLDL remnants."
f. **What is the fate of these newly formed IDLs?**
   (1) **One half travels to the liver and gets absorbed.** (Which two apoproteins mediate IDL uptake by the liver? → **B-100 and E.**) Once absorbed, IDLs are degraded by lysosomes and certain parts (e.g., cholesterol, phospholipids) are recycled.
   (2) **One half continues getting remodeled by HDL and becomes LDLs.** This further remodeling involves two things:
      (a) **HDL removes Apo E**, leaving the resulting LDL with only one remaining apoprotein (B-100). (Why does it make sense for HDL to remove Apo E? → B/c the newly formed LDLs do not need Apo E to bind to tissues. Thus it makes sense for HDL to remove Apo E and recycle it by giving it to someone who will need it—newly forming formed chylomicrons or VLDLs.)
      (b) **HDL adds more cholesterol esters into the core of the IDL,** (which apoprotein mediates transfer of cholesterol esters from HDL? → **CTEP**).
g. **Summary:** TGs synthesized in the liver → VLDLs → peripheral tissues → IDLs (VLDL remnants) → enter liver OR converted to LDLs
h. **Take-home message:** The transport of ENDOGENOUS TGs from the liver to peripheral tissues that need them uses **VLDL**.
4. **ENDOGENOUS TG transport, SCENARIO #2: Your peripheral tissues are working hard and need TGs for energy. Because adipose cells are also a site of TG synthesis, how do the TGs get from adipose to the peripheral tissues that need them?**

    a. TGs synthesized in adipose cells do NOT get released as TGs. There is no lipoprotein "analog" of VLDL that carries TGs from adipose to peripheral tissues.
    b. Instead, adipose cells release fat into circulation by first **degrading TGs and then releasing the resulting FFAs into the blood**. Once in the blood, these FFAs bind to albumin and are carried to peripheral tissues that need them.
    c. **Summary:** TGs synthesized in adipose cells → free fatty acids bound to serum albumin → peripheral tissues
    d. **Take-home message:** Transport of endogenous fatty acids <u>from</u> adipose cells <u>to</u> peripheral tissues or <u>to</u> the liver uses albumin.
  5. **Summary of fat homeostasis in the body**
    a. **If the liver wants to release fats for use by peripheral tissues, how does it do this?** → It packages TGs into VLDLs and releases them.
    b. **If adipose cells want to release fats for use by peripheral tissues, how do they do this?** → They release FFAs into the blood (which bind to serum albumin).
    c. **If a hard-working peripheral cell needs fats from the bloodstream, where do these fats come from and how are they transported to the cell?**
      (1) Exogenous TGs come from the diet and are transported by chylomicrons.
      (2) Endogenous TGs come from the liver and are transported by VLDLs.
      (3) Endogenous fatty acids come from adipose cells and are transported by serum albumin.
    d. **If a peripheral cell has <u>too many</u> TGs or FFAs and wants to get rid of them, what will it do?**
      (1) ↑ Beta oxidation of FFAs, ↓ de novo FFA synthesis
      (2) Release some FFAs into the blood (where they will bind to albumin)
K. **PHOSPHOLIPIDS: TWO FATTY ACIDS AND ONE POLAR GROUP ATTACHED TO A BACKBONE OF GLYCEROL OR SPHINGOSINE**
  1. **The two types of phospholipids are phosphoglycerides and sphingolipids.** Phospholipids that use glycerol as a backbone are called "phosphoglycerides"; those that use sphingosine are called "sphingolipids."

KEY IDEA: The structure of all phospholipids gives them both a hydrophilic (polar) and a hydrophobic (nonpolar) part. This makes them uniquely suited to carry out many physiological roles.
- **Main role of phospholipids?** → Primary structural component of cell membranes. (Remember the often-quoted phrase "phospholipid bilayer" in reference to cell membranes? Well, these are the phospholipids we're talking about.)
- **Other roles of phospholipids?**
- In bile
- As components of lung surfactant
- As components of the outer membrane layer of lipoproteins
- As components of myelin sheaths (sphingomyelin)

2. **Phosphoglycerides**
   a. Use glycerol as a backbone
   b. **All phosphoglycerides contain what molecule as part of their structural foundation?** → Phosphatidic acid
3. **Sphingolipids**
   a. **All sphingolipids contain what molecule as part of their structural foundation?** → Ceramide

What is the point, you ask, of sphingowhatyoumacallits?

> **HPI:** 15-month-old child of **Ashkenazi Jewish** parents presents with **hepatosplenomegaly, lymphadenopathy,** and apparent visual problems. The child is dissimilar to his older sibling, who was able to walk at 15 months of age and say a few basic words (developmental delay). On PE, the above is confirmed as well as a **classic "cherry spot" on the macula,** as in a central retinal artery occlusion on funduscopic examination. A diagnosis is made.
> → Niemann-Pick disease
> - Autosomal-recessive deficiency of **sphingomyelinase** with sphingomyelin accumulation in lysosomes

Everything you need to know about bile
1. **General principles**
   a. Bile is not one thing; it is a nasty-looking mixture of a bunch of stuff:
      (1) Bile salts
      (2) Cholesterol
      (3) Phosphatidylcholine (a.k.a., lecithin) (a phospholipid)
      (4) Bilirubin

> **HUGE KEY IDEA:** There are basically three important concepts regarding bile that you need to know. They will each be discussed further, but here is a brief listing of them:
> 1. Bile contains molecules with both hydrophobic and hydrophilic properties, thus it is able to play an indispensable role in the digestion of fats.
> 2. Your body has very little bile. Therefore, after it is secreted into the duodenum, it must be reabsorbed and "recycled" somehow. This process is called **"enterohepatic circulation."**
> 3. The enterohepatic circulation of bile is ALSO the enterohepatic circulation of bilirubin and of cholesterol. Why? → B/c bilirubin and cholesterol are components of bile.

   b. Bile acids get conjugated with glycine or taurine to become bile salts. Why? → Glycine and taurine help make them more water soluble.
   c. Bile is manufactured in the liver, stored and concentrated in the gallbladder, and secreted through the ampulla of Vader into the duodenum.
   d. Bile is cool because it has both hydrophobic and hydrophilic properties (similar to a detergent). Thus it is able to penetrate globs of fat in the GI tract, divide them into smaller droplets, and then surround these droplets so as to keep them soluble.

KEY IDEA: Your body does not have that much bile. The entire bile acid pool is secreted and reabsorbed two or more times during a single meal!

    e. Rate-limiting step in the formation of bile acids? → 7 alpha-hydroxylase
    f. Inhibitor of bile acid formation? → Cholic acid (inhibits 7 alpha-hydroxylase)
    g. Most potent stimulus of bile secretion? → CCK (signals the presence of fat in GI tract)
    h. Percentage of secreted bile acids reabsorbed each day? → 80% (20% are lost in the feces)
2. **Enterohepatic circulation of bile (you're gonna love this!)**

BIG, MONDO KEY IDEA: The enterohepatic circulation of bile does not just have to do with bile. The enterohepatic circulation of bile is ALSO the enterohepatic circulation of cholesterol and of bilirubin. Bile, cholesterol, and bilirubin get absorbed and secreted together.

M. EVERYTHING YOU NEED TO KNOW ABOUT BILIRUBIN, BUT WERE AFRAID TO ASK

1. **Bilirubin synthesis begins with the death of RBCs.** RBCs get old and are absorbed by the spleen and liver for degradation. The heme group gets broken apart, the Fe_{2+} atom is recycled (i.e., used to synthesize new heme), and the polypeptide chain goes on a fantastic voyage in which it eventually becomes bilirubin.
2. **Fantastic Voyage of the polypeptide heme chain: It gets converted to a bunch of different crud and ultimately ends up as bilirubin, which ends up being secreted as part of bile** (Table 11.2).

TABLE 11.2 Fantastic Voyage: Heme → Bilirubin → Bile

	Site	What Happens
Step 1	Inside a macrophage in the liver or spleen	(1) Heme gets converted to biliverdin and then to bilirubin (heme → biliverdin → bilirubin) (2) Bilirubin from the spleen gets secreted into the bloodstream
Step 2	Bloodstream	Albumin binds bilirubin and takes it to the liver Why must albumin bind bilirubin? → B/c free, unconjugated bilirubin can cross the BBB and is toxic to the CNS
Step 3	Liver	(1) Conjugates bilirubin with glucuronate (2) Releases conjugated bilirubin (bilirubin-diglucuronide) into the bile canaliculi to form part of the bile
Step 4	Gallbladder	Bile is released into the duodenum
Step 5	Small intestine	Bacteria hydrolyze bilirubin-diglucuronide into urobilinogen

Continued

TABLE 11.2 Fantastic Voyage: Heme → Bilirubin → Bile—cont'd

	Site	What Happens
Step 6	Terminal ileum of small intestine	Some urobilinogen gets reabsorbed as part of bile reabsorption. It is then sent to the kidneys where it gets turned into urobilin (a yellow substance) and excreted in the urine. (This is what makes urine yellow.) Review: What makes urine yellow? → Urobilin

a. **What happens to urobilinogen that does not get reabsorbed along with bile in the terminal ileum?** → Enters large intestine and gets converted by bacteria into stercobilin (a brown substance which gives feces its color). Then it gets excreted as feces.

b. Review: What makes feces brown? → Stercobilin (Figure 11.3)

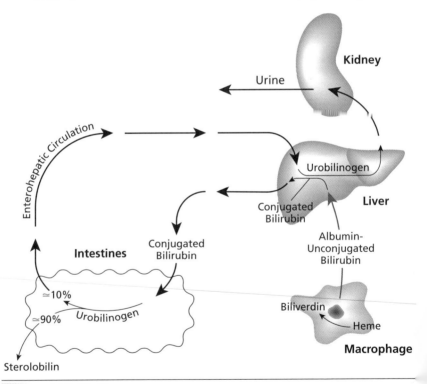

FIGURE 11.3 Bilirubin voyage.

3. **Conjugated (a.k.a., "direct") versus unconjugated (a.k.a., "indirect") bilirubin**
   a. UNconjugated = INdirect = INsoluble
   b. Conjugated = bound to glucuronate = direct = water soluble = able to be excreted in urine
4. **Jaundice: A clinical sign of elevated bilirubin levels in the blood**
   a. What are the classic signs of jaundice? And why? (Table 11.3)

TABLE 11.3 Signs of Jaundice with Explanations

Symptom	Mechanism
↑ Blood [bilirubin]	Blockage of bile secretion (hence bilirubin secretion) causes backflow into the liver bile canaliculi and ultimately into the bloodstream.
Yellow eyes and skin	Some bilirubin in the blood gets turned into urobilin (which is yellow). This can be seen when it lodges in the capillaries of the eyes and skin.
Light-colored stool	Blockage of bile secretion → less bilirubin arriving in the GI tract → less conversion of urobilinogen to stercobilin (a brown substance) → light-colored stools.
Dark-colored urine	Some of the bilirubin in the blood gets turned into stercobilin (which is brown). It is then filtered out of the blood by the kidneys and into the urine.

KEY IDEA: Jaundice is not, in and of itself, a disease! It is a <u>sign</u> of an underlying disorder!

   b. In fact, there are basically three different underlying disorders that can produce jaundice. What could be causing bilirubin in the blood?
   (1) **↑ Production of bilirubin beyond what the liver can conjugate and excrete as bile**
      ex) Massive hemolysis. A surge in the number of RBCs being degraded (as may occur in a patient with sickle cell anemia or malaria) can overwhelm the liver. The result is unconjugated bile spilling into the blood.
   (2) **↓ Excretion of bile so that bile backs up and overflows into the blood**
      ex) Obstructed bile duct. Causes conjugated bile to spill into the blood.
      ex) Liver cell damage (e.g., from cirrhosis or hepatitis). Hepatocytes lose ability to conjugate bilirubin and excrete it as bile. This causes unconjugated bile to spill into the blood.
   (3) **Displacement of unconjugated bilirubin from binding sites on plasma albumin** causes unconjugated bile to spill into the blood.

Clinical correlation. Now that you know everything about jaundice, let's try some cases:

> **HPI:** 18 yo African American woman presents with "stomach" and right upper quadrant pain for many months. Neither the patient nor her family has any history of gallstones. On PE, jaundice is noted on the sclera bilaterally. Blood tests reveal slightly elevated LFTs, increased direct and indirect bilirubin, and a high ratio of coproporphyrin I:III of 5:1 in the urine. Ultrasound rules out gallstones or any definable hepatic processes.
> → Dubin-Johnson syndrome
> - **Note:** Benign autosomal recessive (defective bilirubin excretion)

> **HPI:** 5 yo presents in clinic with mildly yellow skin and sclera, dark-colored urine, and light-colored stool since birth. He has been completely **asymptomatic**, including being **without pruritus** (itchiness). His "yellow" symptoms are exacerbated by stress and colds. Although an incidental finding, a full work-up with labs and MRI show a **conjugated hyperbilirubinemia, normal AST, ALT, albumin, alkaline phosphatase, bile salts, cholesterol, and delayed plasma sulfobromophthalein excretion** after IV injection. The urine shows an **elevated total coproporphyrin excretion** (2.5-5x normal) with 25% being coproporphyrin. A liver biopsy is normal.
> → Rotors syndrome

> **HPI:** 35 yo presents with "yellowing" of skin and eyes, especially when he is stressed. He has no past medical history of any problems with his liver: IV drug use, hepatitis, transfusions, dark urine, etc. On PE, a yellow pallor is noted on his complexion, with scleral icterus and sublingual jaundice. Blood work shows **normal LFTs and a bilirubinemia, which increases after a fast.**
> → Gilbert's disease
> - **Note:** Mainly unconjugated bilirubinemia secondary to defective uptake by liver cells and <u>genetically low levels</u> of UDP glucuronate.

> **HPI:** 2-week-old newborn presents with jaundice that has been worsening since birth. The parents state that the yellow complexion began under tongue, continued from top to bottom. The parents of Jewish descent state that this has been a common problem in the family. Blood work shows a markedly increased unconjugated bilirubin.
> → Crigler-Najjar syndrome
> - **Note:** <u>Complete genetic</u> deficiency of glucuronyl transferase → <u>absence</u> of UDP glucuronate → indirect bilirubinemia → brain damage if > 20 mg/dl.
>   Type I: Severe and autosomal recessive
>   Type II: Autosomal dominant

N. Cholesterol: A crucial molecule that is made from acetyl CoA

1. **Forms of cholesterol in the body**
   a. **Cholesterol versus cholesterol esters**
      (1) Free cholesterol = UNesterified = MORE water soluble
      (2) Cholesterol esters = esterified = LESS water soluble
   b. Intracellular free cholesterol is esterified by ACAT so that it can be stored.
   c. Free cholesterol in the bloodstream is esterified by LCAT so that it becomes insoluble and clumps together, rather than freely inserting itself into nearby cell membranes.
2. **Roles of cholesterol in the body**

> **KEY IDEA:** We think of cholesterol as "bad," but from the body's point of view, cholesterol is a wonderful, precious substance whose homeostasis must be tightly maintained.

3. **Why is cholesterol so important to the body? (two main reasons)**
   a. It is a precursor for synthesis of vitamin D, bile acids, and steroid hormones.
   b. It is a structural component of cell membranes.
4. **Synthesis of cholesterol**
   a. **Location:** Cytoplasm of all cells, but especially the liver
   b. **Metabolic context:**
      (1) You just ate a meal, and circulating insulin is stimulating cholesterol synthesis **(or)**
      (2) Intracellular [cholesterol] has just decreased
   c. **Summary:** Two acetyl CoA → one HMG CoA → one cholesterol
      • Note: Cholesterol synthesis requires NADPH and ATP.
   d. **Nutritional correlation: Cholesterol is NOT needed in the diet.** Your body can synthesize all the cholesterol that it needs using acetyl CoA.
   e. **Rate-limiting enzyme, and reaction, of cholesterol synthesis?**

$$\text{HMG CoA} \xrightarrow{\textit{HMG CoA reductase}} \text{mevalonate}$$

   f. **Regulation of cholesterol synthesis**
      (1) What ↑ **cholesterol synthesis by stimulating HMG CoA reductase?** → Insulin
      (2) What ↓ **cholesterol synthesis?**
         (a) Glucagon (by inhibiting HMG CoA reductase)
         (b) ↑ Intracellular [cholesterol] (by inhibiting transcription of HMG CoA reductase)
         (c) Statin drugs (by inhibiting HMG CoA reductase)

5. Enterohepatic circulation of cholesterol
   a. **What is the enterohepatic circulation of cholesterol?** → It is the enterohepatic circulation of bile (because cholesterol is a part of bile).
   b. **How is cholesterol a part of bile?** → Cholesterol is a precursor for the formation of bile acids. Additionally, some cholesterol is secreted, as is, into the bile.

N. Exogenous and endogenous circulation of cholesterol

> **Key idea:** Both exogenous and endogenous cholesterol transport uses lipoproteins (e.g., chylomicrons, LDLs, HDLs). Lipoproteins have the following structure:
> - Outer membrane = phospholipids, proteins, some free cholesterol
> - Interior core = TGs and cholesterol esters

1. **EXOGENOUS cholesterol transport: Let's say you absorb some cholesterol from the diet. How does it get to the tissues that need it?**
   a. Cholesterol gets **packaged into chylomicrons** and secreted into the lymph system. (Which apoprotein mediates chylomicron secretion? → **B-48.**) Chylomicrons flow through the lymph system and join the bloodstream in the superior vena cava. Sound familiar? Remember: Triglycerides!
   b. **Summary (same as for TGs).** Dietary cholesterol → gut → chylomicrons → lymph → bloodstream → peripheral tissues → chylomicron remnants → liver
   c. **Take-home message: Cholesterol from the diet is packaged into chylomicrons and delivered first to peripheral cells. Any cholesterol that remains in the chylomicron remnants is then absorbed by the liver.**

2. **ENDOGENOUS cholesterol transport: Your peripheral tissues need cholesterol. Since the liver is the major site of cholesterol synthesis, how does cholesterol get from the liver to the peripheral tissues?**
   a. Cholesterol is converted into cholesterol esters and **packaged into LDLs**. LDLs are then released by the liver into the bloodstream.
   b. **Once LDLs are released by the liver into the blood, what are their three fates?**
      (1) Reach peripheral tissues and bind (the intended fate)
      (2) Be absorbed by macrophages
      (3) Break apart and spill their contents into the blood (can occur because of oxidation)
   c. **How do the peripheral tissues bind LDL?** → With LDL receptors that recognize Apoprotein **B-100**
   d. **Once peripheral tissues bind LDL, what three things happen?**
      (1) **Endocytosis** of the LDL
      (2) **Separation** of LDL from its receptor (so the receptor can be recycled)
      (3) **Degradation** of the LDL by a lysosome, causing **release of the cholesterol esters**

e. **Note:** Lysosomal degradation is the end of the line for LDL. It is destroyed. There is no analogous "LDL remnant" that returns to the liver (as is the case with chylomicron remnants or VLDL remnants).

KEY IDEA: Cells regulate the number of LDL receptors they express based on their need for cholesterol.
- If they have plenty of cholesterol, they ↓ the number of LDL receptors (a process called "LDL receptor down-regulation").
- If they lack cholesterol, they ↑ the number of LDL receptors (a process called "LDL receptor up-regulation").

    f. **Once cholesterol released from LDL enters cells, what three intracellular effects does it produce?**
       (1) Inhibits transcription of LDL receptors
       (2) Inhibits activity of HMG CoA reductase (the rate-limiting step in cholesterol synthesis)
       (3) Stimulates activity of ACAT (which esterifies intracellular cholesterol so that it can be stored)
    g. **When cholesterol is used by peripheral cells, what is it used for?**
       (1) All cells: Cholesterol is a structural component of cell membranes
       (2) Gonads and adrenal cells only: Cholesterol is a precursor for steroid hormone synthesis
       (3) Skin cells only: Cholesterol is a precursor for vitamin-D synthesis (process requires sunlight)
    h. **Summary:** Cholesterol synthesized in liver → LDLs → peripheral tissues
    i. **Take-home message: Needy cells receive cholesterol from the liver via LDLs. Once they bind (using Apo B-100), LDLs undergo endocytosis, release their cholesterol, and are destroyed.**
3. **Summary of cholesterol transport to a needy cell.** If a peripheral cell needs cholesterol, where does the cholesterol come from and how is it transported to the cell?
    a. The cell can synthesize its own cholesterol from acetyl CoA.
    b. Exogenous cholesterol comes from the diet, and is transported by chylomicrons.
    c. Endogenous cholesterol comes from the liver and is transported by LDLs.
4. **Summary of cholesterol homeostasis in the body**
    a. **The liver contains a large pool of cholesterol. What are the three sources of this cholesterol?**
       (1) Synthesized by the liver itself
       (2) Absorbed from the diet and transported to the liver via chylomicron remnants
       (3) Synthesized in peripheral tissues and transported to the liver via HDLs
    b. **Peripheral cells also contain cholesterol. What are the three sources of this cholesterol?**
       (1) Synthesized by the cell itself
       (2) Absorbed from the diet and transported to the cell via chylomicrons
       (3) Synthesized in the liver and transported to the cell via LDL

c. **If some cholesterol accidentally gets spilled into the bloodstream, how does the body recycle this precious substance?** → Plasma LCAT esterifies it so that it becomes insoluble. Then HDL scoops it up and takes it to the liver.
d. **Why does it make sense for plasma LCAT to make cholesterol in the blood INsoluble?** → So that it clumps together rather than freely inserting itself into nearby cell membranes.
e. **If the body has <u>too little</u> cholesterol, how does it respond?**
   (1) ↑ LDL receptor synthesis (so you can snag more LDLs from the blood—each of which is loaded with cholesterol from the liver)
   (2) ↑ Intracellular cholesterol synthesis (in all cells, but especially in the liver)
   (3) ↑ Transport of cholesterol from liver to tissues using LDL
   (4) ↓ Secretion of cholesterol as part of bile
f. **If the body has <u>too much</u> cholesterol, how does it respond?**
   (1) ↓ LDL receptor synthesis
   (2) ↓ Intracellular cholesterol synthesis (in all cells, but especially in the liver)
   (3) ↓ Transport of cholesterol from liver to tissues using LDL
   (4) ↑ Secretion of cholesterol as part of bile

5. **Summary and review: Four important functions of HDL**
   a. Transports excess cholesterol from peripheral tissues <u>to</u> the liver (after it is first esterified by LCAT)(Which HDL apoprotein activates LCAT? → A-1.)
   b. Scoops up any free cholesterol floating in the blood and carries it back to the liver (after it is first esterified by LCAT)
   c. Modifies newly released chylomicrons and VLDLs by adding apoproteins C-2 and E to them
      (1) Apo C-2 allows VLDL and chylomicrons to react with lipoprotein lipase on the surfaces of target peripheral cells.
      (2) Apo E mediates uptake of chylomicron and VLDL remnants by the liver.
   d. Remodels chylomicrons and VLDLs after they are hydrolyzed by lipoprotein lipase. Remodeling has four parts:
      (1) Removing part of the cell membrane
      (2) Removing Apo C-2
      (3) Removing any remaining TGs
      (4) Adding cholesterol esters (Which apoprotein mediates transfer of cholesterol esters to VLDLs and chylomicrons? → CTEP.)
6. **Summary: Two common clinical interventions to control cholesterol and how they work** (Table 11.4)

P. **Summary of the apoproteins and their functions** (Table 11.5)

Q. **Summary of lipoproteins and their functions in both TG and cholesterol transport** (Table 11.6)

Table 11.4 Two Clinical Interventions to Control Cholesterol

Intervention	Mechanism
HMG Co-A-reductase inhibitors (e.g., statin drugs) Atorvastatin, pravastatin, simvastatin, lovastatin, youhateboardsastatin, etc.	(1) HMG Co-A reductase is the rate-limiting step of cholesterol synthesis. (2) ↓ Intracellular chol → ↑ LDL receptor synthesis (compensatory response) ↓ ↑ LDLs removed from circulation ↓ ↓ Serum LDL and cholesterol
Inhibition of bile reabsorption in the terminal ileum (e.g., drugs that bind bile acids: cholestyramine resin and colestipol hydrochloride	The body responds to lower levels of bile by ↑ the consumption of cholesterol to synthesize new bile.

Table 11.5 Summary of the Apoproteins and their Functions

Apoprotein	Function
B-100	• Mediates LDL binding to receptors in peripheral tissues and liver for subsequent endocytosis • Mediates VLDL secretion from the liver • Co-mediates uptake of IDL by the liver (together with Apo E)
Apo E	• Co-mediates uptake of chylomicron remnants and IDLs by the liver (chylomicron remnant uptake by liver = Apo E and B-48) (IDL uptake by liver = Apo E and B-100)
Apo B-48	• Mediates chylomicron secretion from GI tract brush border cells into the lymph • Co-mediates uptake of chylomicron remnants by the liver (together with Apo E)
Apo C-2	• Needed by chylomicrons and VLDL to activate lipoprotein lipase on cell surfaces
Apo A-1	• Needed by HDL to activate plasma LCAT as part of removing free cholesterol from the bloodstream
CETP	• Needed by HDL to mediate transfer of cholesterol esters from HDL to VLDL and chylomicrons
A's	• Important structural component of HDL

Table 11.6 Summary of Lipoproteins and their Functions

Lipoprotein	Function	Notes
Chylomicrons	• Transport dietary TGs from the gut to peripheral tissues • Transport dietary cholesterol from the gut to peripheral tissues and the liver	• Secreted by intestinal epithelial cells • Travels in the lymph before joining the bloodstream in the SVC
VLDL	• Transport TGs synthesized in the liver to peripheral tissues	• Excess causes pancreatitis

Continued

Table 11.6 Summary of Lipoproteins and their Functions—cont'd

Lipoprotein	Function	Notes
IDL (a.k.a., VLDL remnant)	• Receives cholesterol esters from HDL and brings them to the liver	• IDLs come from VLDLs. (IDLs are created during VLDL remodeling by HDL) • Think of them as "VLDL remnants" which travel from peripheral tissues to the liver
LDL	• Transport cholesterol synthesized in the liver to peripheral tissues	• Genetic lack of LDL receptors is the basis for familial hypercholesterolemia • What are the 2 sources of LDLs? (1) Synthesized and released by the liver (2) Created during VLDL remodeling by HDL • LDL = Lousy. LDL is bad for the heart because it is packed full of cholesterol being taken to tissues. It is susceptible to being degraded and dumping its contents into the bloodstream.
HDL	• Transports excess cholesterol from peripheral tissues to the liver • Scoops up any free cholesterol floating in the blood and carries it back to the liver • Modifies newly released chylomicrons and VLDLs by adding apoproteins C-2 and E to them • Remodels chylomicrons and VLDLs after they are hydrolyzed by lipoprotein lipase • Remodeling has 4 parts: (1) Removing part of the cell membrane (2) Removing Apo C-2 (3) Removing TGs (4) Adding cholesterol esters	• HDL = Healthy. HDL is good for the heart because it removes free cholesterol floating in blood vessels, thus preventing such cholesterol from becoming part of a plaque

R. **Summary of lipoproteins and the apoproteins they contain** (Table 11.7)

S. **Summary and review of endogenous and exogenous fat and cholesterol transport**
   1. **Important players in TG transport** (Table 11.8)
   2. **Important players in cholesterol transport** (Table 11.9)

T. **Steroid hormone synthesis: An important use of cholesterol**
   1. **General principles**
      a. **What are the five classes of steroid hormones?**

Table 11.7 Lipoproteins and the Functions of their Apoproteins

Lipoprotein	Function of Apoprotein	Name of Apoprotein
Chylomicron	(1) Mediates secretion of chylomicrons from brush border cells into the lymph?	(1) B-48
	(2) Needed to activate lipoprotein lipase on cell surfaces so that the chylomicron can release its contents?	(2) C-2
	(3) Mediates uptake of chylomicron remnants by the liver?	(3) E and B-48
VLDL	(1) Mediates secretion of VLDL from the liver?	(1) B-100
	(2) Needed to activate lipoprotein lipase on cell surfaces so that the VLDL can release its contents?	(2) C-2
IDL (aka, VLDL remnants)	Mediates IDL uptake by the liver?	E and B-100
LDL	Mediates binding of LDL to cell surface receptors for subsequent endocytosis?	B-100
HDL	(1) Activates plasma LCAT as part of removing free cholesterol from the bloodstream?	(1) A-1
	(2) Mediates transfer of cholesterol esters from HDL to VLDL and chylomicrons?	(2) CETP
	(3) Important component of HDL structure?	(3) A's
	(4) Carries around these 2 in order to transfer them to VLDL and chylomicrons?	(4) C-2 and E

Table 11.8 Summary of Triglyceride Transport

• Transports dietary TGs from gut → peripheral tissues?	Chylomicrons
• This enzyme sits on cell surfaces and hydrolyzes chylomicrons to release the stored TGs?	Lipoprotein lipase
• Transports endogenous TG from liver → peripheral tissues?	VLDL
• Transports endogenous FFA from adipose cells → other tissues?	Albumin

Table 11.9 Summary of Cholesterol Transport

• Transports dietary cholesterol from gut → peripheral tissues?	Chylomicrons
• This enzyme sits on cell surfaces and hydrolyzes chylomicrons to release the stored cholesterol?	Lipoprotein lipase
• Transports endogenous cholesterol from liver → peripheral tissues that need it?	LDL
• Transports excess endogenous cholesterol from peripheral tissues → liver?	HDL
• Sits in plasma and esterifies any free cholesterol so that it can be taken up by HDLs?	LCAT

(1) Glucocorticoids (e.g., cortisol)
(2) Mineralocorticoids (e.g., aldosterone)
(3) Androgens
(4) Estrogens
(5) Progestins

b. Steroid hormones, for the most part, do not travel in the blood unbound. They travel bound to albumin or to some other specific carrier protein (e.g., sex hormone–binding globulin).
2. **Mechanism of action of steroid hormones: (four steps)**
   a. Diffuse across cell membrane
   b. Bind to intracellular receptor (either in the cytoplasm or in the nucleus)
   c. Convert receptor to active form
   d. Steroid-receptor complex induces or inhibits transcription of certain genes
3. **Steroid hormone synthesis**
   a. Steroid hormones are synthesized as needed within the gland of origin.
   b. **How?** → Through a series of reactions that use intracellular cholesterol as a starting point
   c. **Summary of steroid hormone synthesis:** Cholesterol → pregnenolone → progesterone → many different pathways. (**Note:** Steroid hormone synthesis requires NADPH and O_2.)
   d. **What is the rate-limiting enzyme, and reaction, of steroid synthesis?**

1. **Testosterone and estrogen production in males and females**
   a. **Males**
      (1) **Where is <u>testosterone</u> made in males?**
         (a) Leydig cells of testes (95%)
         (b) Zona reticularis of the adrenal gland (5%)
      (2) **What hormone stimulates testosterone production in Leydig cells?** → LH
      (3) **How does testosterone synthesis in the gonads differ from the adrenals?** → Adrenals use DHEAS as intermediate; gonads do NOT.
      (4) **Where are <u>estrogens</u> made in males?**
         (a) Peripheral conversion of androgens by aromatase (80% to 90%)
         (b) Conversion of androgens by aromatase in the Sertoli cells of the testes (10% to 20%)
      (5) **What hormone stimulates estrogen production in the Sertoli cells of the testes?** → FSH. **How?** → By ↑ transcription of aromatase
   b. **Females**
      (1) **Where is <u>testosterone</u> made in females?**
         (a) Theca cells of ovaries (50%)
         (b) Zona reticularis of the adrenal gland (50%)
      (2) **What hormone stimulates testosterone production in the theca cells?** → LH
      (3) **How does testosterone synthesis in the gonads differ from the adrenals?** → Adrenals use DHEAS as intermediate; gonads do NOT.
      (4) **Where are <u>estrogens</u> made in females?** → Conversion of androgens by aromatase in the granulosa cells of the ovaries

(5) **What hormone stimulates estrogen production in the granulosa cells of the ovaries?** → FSH. **How?** → By ↑ transcription of aromatase

5. **Quick review of facts you need to know**
   a. Androgens—how many carbons? → 19
   b. Estrogens—how many carbons? → 18
   c. Enzyme that converts testosterone to estradiol? → Aromatase
   d. Type of testosterone that is most potent? → Dihydrotestosterone (DHT)
   e. Enzyme that converts testosterone to DHT (often in peripheral tissues)? → 5-alpha reductase
   f. Used as an index of <u>androgen</u> secretion? → Urine 17-<u>ketosteroids</u>
   g. Used as an index of <u>cortisol</u> secretion? → Urine 17-<u>OH steroids</u>

Some clinical cases:

HPI: 15 yo adolescent presents with **lack of pubic and axillary hair, amenorrhea,** and **lack of breast development.** On PE, blood pressure is noted to be quite high **175/110.** Funduscopic examination reveals no abnormalities. The rest of the examination is normal except for the complete lack of pubic and axillary hair. A Chem 7 reveals a sodium of 160 **(hypernatremia)** and potassium of 2.0 **(hypokalemia).** A blood gas shows **alkalosis.**
→ 17-alpha-hydroxylase deficiency
- With the lack of this enzyme in the steroid synthesis pathway, increased corticosteroids are produced. Both females and males lack necessary sex hormones for secondary sexual characteristics.

HPI: Newborn is lethargic and receives a decreased APGAR, specifically for muscle tone. The ob/gyn is unable to pronounce the sex of the child at birth. The "labia" appear fused, and the "clitoris" or "penis" is indistinguishable. Blood draw reveals a potassium of 6 **(hyperkalemia)** and slight hyponatremia. Further testing is done and shows an increase in **17-alpha-OH progesterone, urinary ketosteroids, and ACTH.** A karyotype is done to ascertain sex: 46 XX.
→ Congenital adrenal hyperplasia/21-hydroxylase deficiency
- "Too Manly" syndrome when an enzyme deficiency results in decreased cortisol production. Feedback loops go haywire and hyperplasia of the adrenals gives rise to excess androgen production causing female pseudohermaphroditism and male macrophallia.

HPI: 2 yo presents with **microphallus** (small penis—but how would you know?), **cryptorchidism** ("hidden testis"), and **hypospadias** (urinating through an orifice anywhere along the developmental canal). PE confirms the complaints of the parents. Lab testing of sex hormones shows a markedly **decreased DHT level with normal testosterone**.
→ 5-alpha reductase deficiency
- **Note:** Cryptorchidism predisposes to testicular cancer not because of some physiologic property of being stuck up there, but rather the probability of detection is much lower! DHT is a potent activator of male secondary sex characteristics.

Chapter 12
Protein and Nitrogen Metabolism

Note: "NH_3" and "NH_{4+}" are used interchangeably in this chapter to refer to the alpha amino group on an amino acid. Consider the terms to be synonymous.

A. General principles
1. **In metabolism, it is helpful to think of an amino acid as consisting of two parts—a carbon skeleton and an NH_3 group.**
   a. Amino acid = carbon skeleton + NH_3

KEY IDEA: Amino acids can be converted into carbon skeletons and vice versa (by removal or addition of an NH_3 group, respectively) (Figure 12.1).

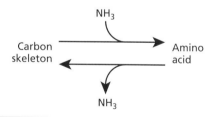

Figure 12.1 Interconversions between amino acids and their carbon skeletons.

   b. Here are some common amino acids and their corresponding carbon skeletons. You must know these.

Amino acid	↔	**Carbon skeleton**
Glutamate	↔	Alpha-KG
Alanine	↔	Pyruvate
Aspartate	↔	Oxaloacetate (OAA)

2. **Why is adequate dietary protein intake necessary?** → B/c amino acids are needed to synthesize important stuff, such as:
   a. Different **cellular proteins** (e.g., cell membrane receptors, albumin)
   b. **Neurotransmitters and hormones** (e.g., catecholamines, peptide hormones)

c. **Purines and pyrimidines** (for use as nucleotides)
d. **Porphyrins** (e.g., heme, which is used in hemoglobin, myoglobin, and cytochrome)

> **KEY IDEA:** If dietary protein intake is not plentiful enough, or does not include the essential amino acids, how does the body get the amino acids that it needs to carry out the above processes? → By breaking down skeletal muscle. After skeletal muscle supplies are exhausted, the body begins to break down the proteins of organs and other tissues. This is how starvation kills a person.

Let us try two similar cases that illustrate what happens when the body starts breaking down proteins from muscles and organs.

> **HPI:** 3 yo child found abandoned in Kashmir presents with **increased abdominal girth** and failure to thrive. On PE, height and weight are below the 5th percentile. **Bilateral pitting edema and ascites** are noted. CBC shows **anemia**. A Chem 7 (electrolyte panel) has irregularities. A **low albumin** count is noted. UTZ (ultrasound) of the abdomen reveals an enlarged and apparently "**fatty liver.**" What is the diagnosis?
> → Kwashiorkor

> **HPI:** 3 yo child from the streets of Kabul presents with **diarrhea, dehydration, and behavioral changes**. On PE, dry, **loose skin and brittle hair** is found. The child is less than the 5th percentile for height and weight. In addition to the growth retardation (**weight** greater than height), there is progressive wasting of subcutaneous fat and muscle. Upon treatment with a high-calorie, protein-rich diet, the patient's condition markedly improves. Name this nasty little condition.
> → Marasmus

3. **Nutritional correlation:** Note the following relationships between daily caloric intake and daily protein requirement:
   a. As total energy intake ↑ the daily need for protein ↓.
      (1) Why? B/c you have so many incoming food substrates, the need to catabolize body proteins as a means of generating energy decreases. Therefore the required intake of proteins to replace those lost via catabolism decreases.
      - Basically, with an energy-plentiful diet, the body prefers using dietary fats and carbohydrates, rather than its own proteins, for energy.
   b. As total energy intake ↓ (e.g., dieting) the daily need for protein ↑.
      (1) Why? Your body responds to ↓ caloric intake by mobilizing its stored fuel substrates. In other words, the use of TG, glycogen, and body proteins for energy ↑. B/c of increased catabolism of body proteins to generate energy, the required intake of proteins replaces those lost by catabolism increases.

4. **Within a cell, proteins are constantly being degraded and synthesized.**
   a. **When is protein <u>degradation</u> most elevated?** → During periods of intense exercise, or fasting.
   b. **Why?** → B/c proteins can be used to generate energy.
   c. **How do proteins help generate energy?** → Via a two-step process:
      (1) The protein gets <u>broken down into its constitutive amino acids</u>.
      (2) The <u>amino acids are deaminated</u> (i.e., NH_{4+} groups are removed), and the resulting <u>carbon skeletons are fed into the TCA cycle</u> (all cells) **OR** <u>into gluconeogenesis</u> (liver and kidney cells only).
   d. **When is protein <u>synthesis</u> most elevated?**
      (1) After a meal (insulin stimulates protein synthesis)
      (2) During sleep (HGH released during deep sleep stimulates protein synthesis)
      (3) During a growth spurt (HGH and androgen release during growth spurt stimulates protein synthesis)

<u>**HUGE KEY IDEA:**</u> **You must understand the structure and roles of alpha-KG, glutamate, and glutamine.**
- Gluta**mate** and gluta**mine** are both NH_3 carriers.
- Gluta**MATE** has **ONE** NH_3 group. → (When glutaMATE donates its NH_3, it becomes alpha-KG.)
- Gluta**MINE** has **TWO** NH_3 groups. → (When glutaMINE donates its first NH_3, it becomes glutaMATE.)
- **Summary:** (Figure 12.2)

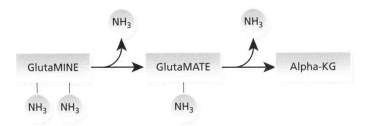

FIGURE 12.2 Glutamate versus glutamine versus alpha kG.

KEY IDEA: There are two main reactions that occur in amino acid metabolism: Transamination and deamination. You must understand them. (These reactions will be discussed later.)

B. **WAYS TO METABOLICALLY CLASSIFY AMINO ACIDS: (1) GLUCOGENIC, KETOGENIC, OR BOTH AND (2) ESSENTIAL OR NONESSENTIAL**
   1. **Amino acids that are glucogenic, ketogenic, or both:**
      a. **Ketogenic amino acids** give rise to ketone bodies (i.e., are converted to acetoacetate, acetone, or beta-hydroxybutyrate).

b. **Glucogenic amino acids** give rise to glucose and glycogen by entering gluconeogenesis.
c. **These distinctions allow for three categories:**
   (1) Exclusively glucogenic
   (2) Glucogenic AND ketogenic
   (3) Exclusively ketogenic (Table 12.1)

TABLE 12.1 **Metabolic Classifications of Amino Acids**

	Nonessential	Essential
Glucogenic	Alanine Asparagine Aspartate Cysteine Glutamine Glutamate Glycine Proline Serine	Arginine Histidine Methionine Threonine Valine
Glucogenic and ketogenic	Tyrosine	Isoleucine Phenylalanine Tryptophan
Ketogenic	Leucine	Lysine

d. **Review.** → **What are the only two exclusively ketogenic amino acids?**
   (1) Leucine
   (2) Lysine
2. **Essential versus nonessential amino acids**
   a. **Essential amino acids:** The body cannot synthesize them, thus they must be obtained from the diet.
   b. **Nonessential amino acids:** The body can synthesize them (using carbon skeletons from the TCA cycle and NH_3 from glutamate).
   c. Pneumonic for the essential amino acids: PVT TIM HALL (Figure 12.3)

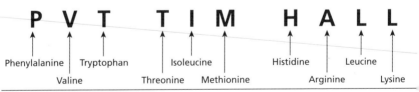

FIGURE 12.3 The essential amino acids.

C. Transamination: Transfer of NH_3 between an amino acid and alpha-KG

1. **Location of transamination:** Mitochondrial matrix of all cells, but especially the liver and kidney
   a. **Clinical correlation:** The appearance of certain aminotransferase enzymes in the <u>blood</u> (because they should be <u>inside</u> the cell) is indicative of tissue damage.
      ex) AST > ALT elevation in alcoholics/cirrhosis
      ex) ALT > AST in other forms of liver damage

KEY IDEA: Transamination can occur in both directions! This allows it to participate in <u>both</u> synthesis <u>and</u> degradation of amino acids.
- As part of amino acid synthesis (Figure 12.4)
- As part of amino acid degradation (Figure 12.5)

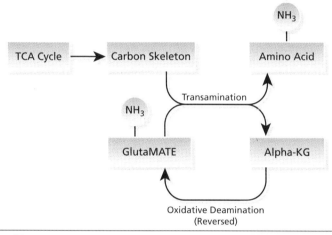

FIGURE 12.4 Transamination and deamination in amino acid synthesis.

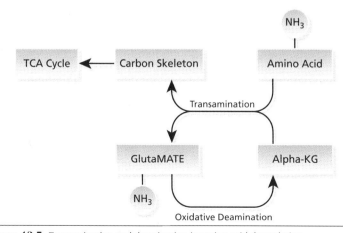

FIGURE 12.5 Transamination and deamination in amino acid degradation.

b. **Nutritional correlation:** All transamination reactions require a derivative of vitamin B_6 (pyridoxal phosphate) to act as a coenzyme along with the particular aminotransferase involved.

> **KEY IDEA:** Notice how glutamate/alpha KG is used an NH_3-carrying intermediate. That is basically its job—to carry NH_3 around!

2. Who cares about transamination?
   a. **It allows for interconversions between different amino acids.** Basically, transamination allows the body to remove NH_3 from one amino acid and transfer it to another, using glutamate as the NH_3-carrying intermediate (Figure 12.6).

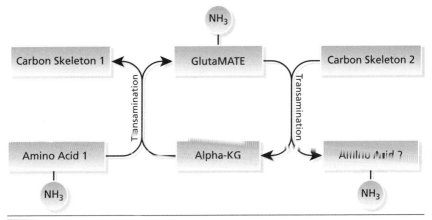

FIGURE 12.6 Interconversions between different amino acids.

   b. **It allows muscle cells to get rid of NH_3 by converting pyruvate to alanine** (which is then sent to the liver). Export of NH_3 by converting pyruvate to alanine, and then sending alanine to the liver, is part of the glucose/alanine cycle (Figure 12.7).
   c. **It works together with oxidative deamination to maintain amino acid homeostasis.** (More on this topic later.)
   d. **It yields carbon skeletons, which can be used for many different purposes.**
      (1) Different amino acids yield different carbon skeletons.
         (a) Some yield pyruvate.
         (b) Some yield acetyl CoA.
         (c) Some yield succinyl CoA.

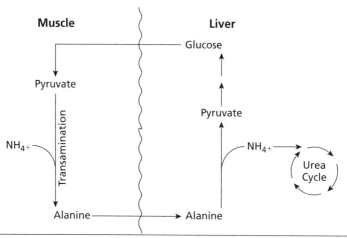

FIGURE 12.7 Glucose-alanine cycle.

(2) Each of these carbon skeletons has its own unique fates within a cell. (See Chapter 8 for details about the fates of pyruvate and acetyl CoA.)

D. OXIDATIVE DEAMINATION: REMOVAL OF NH_3 FROM GLUTAMATE TO YIELD ALPHA-KG

1. **Location of oxidative deamination:** Mitochondrial matrix of all cells, but especially the liver and kidney

> **KEY IDEA:** The oxidative deamination <u>can be reversed!</u> This allows it to participate in <u>both</u> synthesis <u>and</u> degradation of amino acids.
> - As part of amino acid synthesis: (see Figure 12.4)
> - As part of amino acid degradation: (see Figure 12.5)

2. **What happens to the NH_{4+} that gets released from oxidative deamination?**
   a. In **kidney** cells, NH_{4+} gets secreted into the urine.
   b. In **muscles**, NH_{4+} gets converted to alanine, which is then sent to the liver.
   c. In **liver** cells, NH_{4+} enters the urea cycle. The resulting urea is sent to the kidneys.
   d. In all other tissues, NH_{4+} gets converted to glutamine, which is released into the bloodstream (Figure 12.8).
      - Circulating glutamine is then absorbed by either the liver or the kidney.

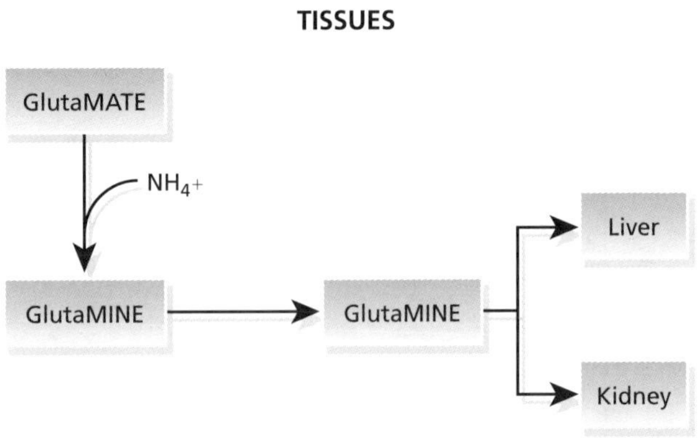

TISSUES

```
GlutaMATE
    |
    | — NH₄⁺
    ↓
GlutaMINE  →  GlutaMINE  →  Liver
                         →  Kidney
```

FIGURE 12.8 Most tissues get rid of NH_4^+ by creating glutamine.

E. THE GLUTAMATE/GLUTAMINE CYCLE: A WAY FOR TISSUES TO SEND NH_4^+ TO THE LIVER OR KIDNEYS (Figure 12.9)

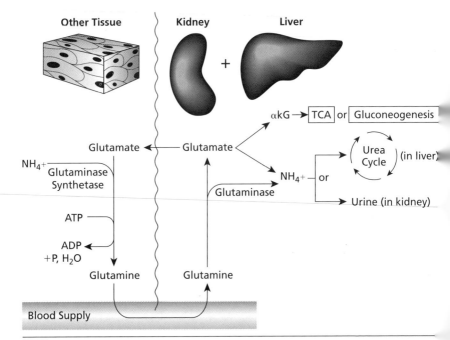

FIGURE 12.9 Glutamate/glutamine cycle.

1. **Here is how the glutamate/glutamine cycle works:**
   a. **In all tissues besides the liver and kidneys**
      (1) Oxidative deamination produces NH_{4+}.
      (2) Cells add an NH_{4+} to gluta**mate** to create gluta**mine** (enzyme = glutamine synthetase).
      (3) Cells release gluta**mine** into the blood; it gets absorbed by either the liver or the kidney.
   b. **In the liver**
      (1) Liver takes up gluta**mine** from the blood.
      (2) Liver removes an NH_{4+} from gluta**mine** to recreate gluta**mate** (enzyme = glutaminase). The NH_{4+} enters the urea cycle.
      (3) Liver either sends gluta**mate** back to the peripheral tissues **or** deaminates it into alpha-KG and NH_{4+}.
         (a) The NH_{4+} enters the urea cycle.
         (b) The alpha-KG enters TCA or gluconeogenesis.
   c. **In the kidney**
      (1) Kidney takes up gluta**mine** from the blood.
      (2) Kidney removes an NH_{4+} from gluta**mine** to recreate gluta**mate** (enzyme = glutaminase). The NH_{4+} is excreted into the urine.
      (3) Kidney either sends gluta**mate** back to the peripheral tissues **or** deaminates it into alpha-KG and NH_{4+}.
         (a) The NH_{4+} is excreted into the urine.
         (b) The alpha-KG enters TCA or gluconeogenesis.
   d. **Renal physiology flashbacks (You know you love these!)**
      (1) **How, exactly, is the NH_{4+} excreted into the nephron? (two ways)**
         (a) The NH_{4+} substitutes for H^+ on the Na^+/H^+ exchanger (or)
         (b) NH_{4+} is converted to NH_3, which can then passively diffuse across the cell membrane.
      (2) **After NH_{4+} is excreted into the tubule, what does it do there?**
         → It works together with phosphate to provide a buffer system for H^+.

F. **THE GLUCOSE/ALANINE CYCLE: A WAY FOR MUSCLE CELLS TO GET RID OF NH_3, AND FOR LIVER CELLS TO SUPPLY MUSCLES WITH FRESH GLUCOSE DURING PERIODS OF STRENUOUS EXERCISE** (see Figure 12.7)

   1. **The glucose/alanine cycle is only used by muscle tissues.** (However, muscle tissues use both the glucose/alanine cycle and the glutamate/glutamine cycle.)
   2. **Basic idea of the glucose/alanine cycle:** Muscles export alanine to the liver—as a means of getting rid of NH_{4+}—and get glucose in return.
   3. **Here is how the glucose/alanine cycle works:**
      a. **In the muscle**
         (1) Muscle adds an NH_{4+} to pyruvate to create alanine.
         (2) Muscle sends alanine to the liver.

b. **In the liver**
   (1) Liver takes up alanine from the blood.
   (2) Liver removes an NH_{4+} from alanine to recreate pyruvate.
      - The NH_{4+} enters the urea cycle
      - Pyruvate enters gluconeogenesis
   (3) Liver sends fresh glucose (from gluconeogenesis) back to the muscle.

G. REVIEW OF AMMONIA (NH_{4+}) METABOLISM: IT IS A DEADLY TOXIN THAT CAN CROSS THE BBB AND HARM THE CNS. IT MUST NOT BE ALLOWED TO ACCUMULATE IN THE BLOOD
   1. **Where does most of the body's ammonia originate?** → Degradation of excess amino acids
      a. Three other sources:
         (1) Amines in the diet
         (2) Bacterial production in gut
         (3) Catabolism of purines and pyrimidines
   2. **Because ammonia is toxic to the CNS, what is the body's strategy to deal with it?** → A two-part strategy:
      a. Excrete NH_{4+} from the kidney.
      b. In the process of transporting ammonia to the kidney, <u>never</u> let free NH_{4+} travel in the blood.
         (1) Thus, ammonia traveling <u>from peripheral tissues to the liver</u> travels as <u>alanine</u> or <u>glutamine</u>.
         (2) And ammonia traveling <u>from the liver to the kidney</u> travels as <u>urea</u>.

> **KEY IDEA:** Where is ammonia found and what is the role of each tissue in ammonia metabolism? (Table 12.2)

TABLE 12.2 Summary: Roles of Different Tissues in Ammonia Metabolism

Tissue in Which Ammonia Is Present	Role of Tissue in Ammonia Metabolism
Muscles	• Converts ammonia to alanine or glutamine; sends these to the liver and kidney
Other peripheral tissues	• Converts ammonia to glutamine; sends these to the liver and kidney
Liver	• Receives incoming glutamine and alanine from the blood; removes the NH_{4+} from them; sends the NH_{4+} to the urea cycle; sends urea to the kidneys • NH_{4+} produced within the liver gets converted to urea; urea gets sent to the kidneys
Kidneys	• Receives incoming glutamine from peripheral tissues; removes the NH_{4+} from it; excretes NH_{4+} into the urine • Receives incoming urea from the liver; removes the NH_{4+} from it; excretes NH_{4+} into the urine • Excretes NH_{4+} produced within the kidney directly into the urine

3. Let us review. In what form does ammonia travel in the blood?
   a. From muscle to the liver? → As alanine or glutamine
   b. From other peripheral tissues to either the liver or kidney?
      → As glutamine
   c. From liver to the kidney? → As urea
   d. **So, why is ammonia never seen in the blood?**
      (1) Because it is traveling as glutaMINE or alanine
      (2) Because it is traveling as urea

HPI: 33 yo woman was in an MVA (motor vehicle accident). She had lost significant amount of blood through hemorrhage of the splenic artery. A splenectomy was done (spleen removed) and on postoperative day one, she complained of thirst and dizziness. Decreased urine output and orthostatic hypotension were noted on PE. Her BUN came back at a level of 50 and the creatinine was 5, both very high. What is your diagnosis?
→ Acute renal failure secondary to prerenal hypovolemia and fluid loss
- The kidneys start to fail and there is subsequent hyperammonemia. How does kidney failure cause hyperammonemia? Something interesting can happen to urea on its way to the kidney—it can diffuse out of the blood and into the small intestine; and the more urea traveling in the blood, the greater the rate of diffusion into the small intestine.
- **Who cares?** Certain bacteria in the small intestine can cleave urea and cause release of NH_3 ☹
- **As I said, who cares?**
  1. This is one reason why patients with kidney failure have hyperammonemia!
     - Failing kidney → ↓ renal metabolism of urea → ↑ blood (urea) → ↑ diffusion of urea into small intestine → ↑ cleavage of urea by bacteria → ↑ release of NH_3 → hyperammonemia
  2. This is the reason why **antibiotics** are given to patients with renal failure to treat their hyperammonemia.
- **As I said, who cares?** → Hmmmm. I think you should consider a career change.

H. TRANSAMINATION AND OXIDATIVE DEAMINATION **WORK TOGETHER** IN BOTH DEGRADATION AND SYNTHESIS TO MAINTAIN AMINO ACID HOMEOSTASIS
  1. **Scenario #1: There are too many amino acids in the cell.**
     a. **Response by the cell: Degradation.** Transamination and oxidative deamination both help degrade excess amino acids into carbon skeletons that enter the TCA cycle.
     b. **Pathway of amino acid degradation:** Breaking an amino acid into a carbon skeleton and NH_3 (Figure 12.10)
        (1) **Step 1: Transamination of the amino acid** to yield glutamate and a carbon skeleton (**amino acid** + alpha-KG → carbon skeleton + **glutamate**). The carbon skeleton gets sent to the TCA cycle.

(2) **Step 2: Oxidative deamination of glutamate** to yield alpha-KG and NH_3 (**glutamate → alpha-KG** + NH_3)
(3) **Step 3: Send the alpha-KG to the TCA cycle**
2. **Scenario #2: There are <u>too few</u> amino acids in the cell.**
   a. **Response by the cell: Synthesis.** Transamination and oxidative deamination <u>both</u> help synthesize amino acids from carbon skeletons removed from TCA cycle.
   b. **Pathway of amino acid synthesis:** Creating an amino acid from a carbon skeleton and NH_3 (Figure 12.11)
      (1) **Step 1: Remove alpha-KG from the TCA cycle**
      (2) **Step 2: Reverse reaction of oxidative deamination** to yield glutamate (alpha KG + NH_3 → glutamate)
      (3) **Step 3: Transamination of glutamate** to yield an amino acid (<u>glutamate</u> + carbon skeleton → alpha-KG + **amino acid**)

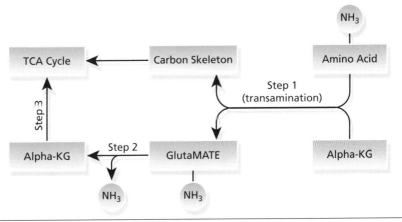

FIGURE 12.10 Amino acid degradation pathway.

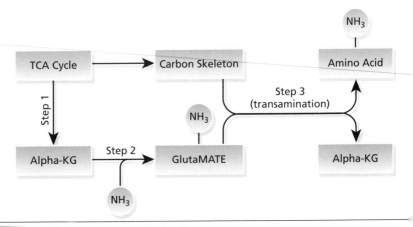

FIGURE 12.11 Amino acid synthesis pathway.

3. **How is the balance between degradation and synthesis of amino acids regulated? (two ways)**
   a. **By the availability of the respective substrates**
      ex) If amino acid levels are high, then the net flux shifts toward degradation.
   b. **By the energy level within the cell (i.e., the levels of ADP/GDP and ATP/GTP)**
      (1) <u>Low-energy levels</u> within the cell are indicated by ↑ [ADP] and ↑ [GDP], which cause net flow in the direction of <u>degradation</u> by acting on glutamate dehydrogenase. **Why does this make sense?** → B/c ↑ degradation = ↑ carbon skeletons entering TCA and generating energy = reversed of low-energy situation.
      (2) <u>High-energy levels</u> within a cell are indicated by ↑ [ATP] and ↑ [GTP], which cause net flow in the direction of <u>synthesis</u> by acting on glutamate dehydrogenase. **Why does this make sense?** → B/c if a cell has plenty of energy it does not need to degrade its amino acids to generate more.

I. **UREA CYCLE: THE LIVER'S WAY OF CONVERTING AMMONIA INTO SOMETHING NONTOXIC AND EXCRETABLE—UREA**
   1. **Net reaction of the urea cycle:** CO_2 + NH_4 + aspartate → fumarate + urea
      a. **What three things enter the urea cycle and where do they come from?**
         (1) CO_2 (from cellular metabolic reactions)
         (2) NH_{4+} (mostly from degradation of amino acids)
         (3) Aspartate (from transamination of OAA)
      b. **What two things leave the urea cycle and where do they go?**
         (1) Fumarate (enters TCA cycle or gluconeogenesis)
         (2) Urea (excreted into the blood and sent to kidneys)
   2. **The urea cycle requires energy. How much?** → Four ATP per molecule of urea produced.
   3. Although it is a cycle, think of the urea cycle as beginning and ending with ornithine.
   4. Rate-limiting enzyme, and reaction, of the urea cycle?

   $$CO_2 + NH_{4+} \xrightarrow[\text{(mitrochondrial matrix)}]{\textit{Carbamoyl phosphate synthetase 1}} \text{carbamoyl phosphate}$$

   5. **What ↑ the rate of the urea cycle by stimulating carbamoyl phosphate synthetase 1?** → N-acetylglutamate (which is most commonly synthesized in response to a protein-rich meal)
   6. <u>Where</u> **are the first two reactions of the urea cycle?** → Mitochondrial matrix of the liver
      a. Reaction 1. CO_2 + NH_{4+} * Carbamoyl Phosphate
      b. Reaction 2. Carbamoyl Phosphate + Ornithine → Citrulline
   7. <u>Where</u> **are the rest of the reactions?** → Cytoplasm of the liver

KEY IDEA: What are the two ways that nitrogen/ammonia can enter the urea cycle?
1. As carbamoyl phosphate
2. As aspartate

7. **In what two ways is the urea cycle connected to the TCA cycle?**
    a. **OAA** (a carbon skeleton in the TCA cycle) can be converted to aspartate (its corresponding amino acid). Aspartate can then enter the urea cycle.
    b. The urea cycle produces **fumarate**, which can enter the TCA cycle.

Chapter 13
Nucleotide Metabolism

A. General principles

1. **N base versus nucleoside versus nucleotide**
   N base (a purine or pyrimidine)
   ↓ (add a pentose sugar—either ribose or deoxyribose)

   NucleoSIDE (N base + pentose sugar)
   ↓ (phosphorylate)

   NucleoTIDE (N base + pentose sugar + at least one phosphate)

2. **Why are nucleotides important?**
   a. Substrates for **DNA synthesis** (i.e., replication) dATP, dGTP, dTTP, dCTP
   b. Substrates for **RNA synthesis** (i.e., transcription) ATP, GTP, UTP, CTP
   c. **Structural components of essential enzymes,** such as CoA, NAD, FAD, $NADP^+$
   d. **"Energy currency"** in the cell (e.g., ATP, GTP)
   e. **Regulatory molecules** that inhibit key metabolic pathways
      ex) Cyclic AMP, cyclic GMP
      ex) ATP, GTP, ADP, GDP
3. **Where do the N bases originate?**
   a. Diet
   b. De novo synthesis
   c. Recycled from degraded RNA/DNA (via salvage pathways)

Key idea: Ribonucleotides can be converted to deoxyribonucleotides.
ex) ADP, GDP, CDP, UDP → dADP, dGDP, dCDP, dUDP
- Enzyme = ribonucleotide reductase

4. **Clinical correlation → How does the drug hydroxyurea inhibit nucleotide synthesis?**
   → It inhibits the conversion of ribonucleotides to deoxyribonucleotides by inhibiting ribonucleotide reductase (RR).

Why do we care?
→ By inhibiting the synthesis of deoxyribonucleotides, hydroxyurea can act as an anticancer agent (stopping rapidly dividing cells), and as an anti-HIV drug in conjunction with antiviral DNA analogs (inhibiting the synthesis of nucleotides forces HIV to incorporate chain-terminating DNA analogs).

KEY IDEA: All of the nucleotide reaction pathways occur in the **cytoplasm of all cells.**

B. **THE HMP SHUNT: A WAY FOR CELLS TO DIVERT GLUCOSE-6-P FROM GLYCOLYSIS TO MAKE NADPH AND RIBOSE-5-PHOSPHATE (WHICH IS THE KEY PRECURSOR MOLECULE OF NUCLEOTIDE SYNTHESIS)**
  1. **Where does it occur?** → Cytosol of all cells
  2. **Reaction summary:** Glucose-6-P → ribose-5-P + 2 NADPH
     a. **Rate-limiting enzyme and reaction of the HMO shunt?**

     $$\text{glucose-6-P} \xrightarrow{\text{glucose-6-P dehydrogenase}} \text{6-P-gluconolactone}$$

KEY IDEA: What coenzyme is required by the HMP shunt? → TPP

  b. **How many NADPH are produced for each glucose-6-P that enters?**
     → 2
  c. **Where does the glucose-6-phosphate originate?** → Glycolysis or glycogen
  d. **Where does the ribose-5-P go?** → Enters nucleotide synthesis pathway
  e. **What are the nonoxidative, reversible reactions?** → Transketolase and transaldolase
  f. **What is the oxidative, irreversible reaction?** → Glucose-6-phosphate dehydrogenase
  3. **Let's say there is no need to produce nucleotides at the time (i.e., the cell is not doing replication or transcription)? What happens to flux through the HMP shunt?**
     → It decreases.
     - Also, any substrates within the pathway can be converted to ribose-5-phosphate, which can then be converted to an intermediate of glycolysis.
     a. **What are the two sites at which HMP shunt products can enter glycolysis?**
        (1) As fructose-6-P
        (2) As glyceraldehyde-3-P

KEY IDEA: Why is the HMP shunt so important? (three main reasons)
1. **It creates NADPH, which is necessary for six crucial physiologic reactions.** Each of these processes requires NADPH:
   a. Steroid hormone biosynthesis
   b. De novo fatty acid synthesis
   c. Nucleotide synthesis
   d. Regeneration of reduced glutathione (a crucial antioxidant in RBCs)
   e. Cytochrome P450 function in the liver (which is responsible for detoxifying drugs and other foreign compounds)
   f. Phagocytosis of foreign bodies by white blood cells (NADPH is involved in the "respiratory burst")
   Is there an alternative pathway to generate NADPH if the HMP shunt is not working? → Yes. NADPH-dependent malate dehydrogenase
2. **It creates ribose-5-phosphate—a required precursor for nucleotide synthesis.**
3. **It provides a way for 5-carbon sugars absorbed from the diet to enter glycolysis** (by converting them to fructose–6-P or glyceraldehyde-3-P).
4. The HMP shunt is particularly important to the following tissues (Table 13.1) **for the following reasons:**

TABLE 13.1

Tissue	Why Is the HMP Shunt Important?
Liver, mammary glands, adipose cells, and kidneys	These are the sites of de novo fatty acid synthesis (and NADPH is needed for de novo synthesis of fatty acids)
Liver	NADPH is necessary for the reactions catalyzed by cytochrome p450 enzymes
Kidney (adrenal cortex), testes, ovaries	These are the sites of steroid hormone synthesis (and NADPH is needed for synthesis of steroid hormones)
Red blood cells	NADPH is needed to regenerate the antioxidant glutathione
White blood cells	NADPH is necessary for reactions that are part of the phagocytotic destruction of foreign bodies

Clinical correlation: What is the most common enzyme deficiency in the world?

> **HPI:** 45 yo African-American man presents with a urinary infection. He is treated with Bactrim (a sulfa drug). One week later, he returns complaining of malaise, "jaundice," and fever-chills-nausea. On PE, he is tachycardic and pale. Blood tests show a low hematocrit and an indirect bilirubinemia. (You should know the significance of this by now, eh?) A peripheral smear shows spherocytes, Heinz bodies, and low enzyme levels of what enzyme?
> → Glucose-6-phosphate (the most common enzyme deficiency in the world).
> - X-linked recessive disorder → RBC sensitivity to oxidant damage and hemolysis.

Continued

So the guy doesn't have much glucose-6-phosphate. Why is this a problem? → G6PD deficiency impairs the body's ability to use the HMP shunt. Therefore it decreases synthesis of NADPH and, as a result, it impairs the six important physiologic processes that require NADPH.

What is the most damaging effect of G6PD deficiency? Hemolysis of RBCs because of an impaired ability to generate NADPH. (NADPH is needed to regenerate glutathione [which is a crucial antioxidant that protects RBCs].)

Summary: G6PD deficiency → ↓ flux through HMP shunt → ↓ NADPH → ↓ antioxident protection of RBCs.

What is responsible for the reduced life span of some G6PD deficiency patients? → Chronic hemolytic anemia resulting from hemolysis of RBCs

What are some things that can precipitate a hemolytic crisis? → Any oxidant stress on the RBCs. *exs)* Certain drugs, infection, fava beans!

Clinical correlation → Lack of flux through the HMP shunt can lead to this syndrome:

HPI: 42 yo alcoholic who has been stuck on a sea-going vessel for 6 months complains of malaise, shortness of breath, extremity swelling, and **"foot drop."** His diet has consisted of crackers, water, and gin for the last several months. PE reveals **nystagmus**. MSE (mental status examination) is abnormal with a score of 24. Extremities are warm, 2+ **pitting edema**. Lungs have bilateral rales in the lower lobes. **Cardiomegaly** is apparent by the shift in the PMI (point of maximal impulse). What is the diagnosis?
→ Thiamine deficiency: Wernicke-Korsakoff syndrome.
- Lack of thiamine → lack of TPP → decreased flow through HMP shunt → decreased NADPH → decreased nucleotide synthesis.
- Patient has wet beriberi (cardiac dysfunction) and dry beriberi (neuropathy).
- Always treat an alcoholic with thiamine before glucose, or else encephalopathy will ensue!

Summary: What are two causes of a ↓ in flux through the HMP shunt?
1. Lack of TPP (thiamine pyrophosphate)
2. G6PD deficiency

C. PURINES—DE NOVO PURINE (AMP, GMP) SYNTHESIS: BUILDING A NUCLEOTIDE FROM SCRATCH

1. **Summary of de novo purine synthesis pathway:**

ribose-5-phosphate → PRPP → 5-phosphoribosylamine → IMP ⎰→ AMP
⎱→ GMP

2. **Primary end-product?** → IMP
3. **What is the problem with de novo pathways?** → They use a lot of energy.
4. **What is an important source of carbon atoms in de novo purine synthesis?** → N-formyl THF (tetrahydrofolate)
5. **What is an important source of N atoms for de novo purine synthesis?**
   → Glutamine, aspartate, glycine
6. **Two key reactions:**
   a. **The rate-limiting step:**

   $$\text{ribose-5-phosphate} \xrightarrow{\textit{PRPP synthesis}} \text{PRPP}$$

   (1) Stimulated by? → Phosphate levels
   (2) Inhibited by? → Downstream end-products—IMP, GMP, AMP
   b. **The committed step:**

   $$\text{PRPP} \xrightarrow{\textit{PRPP aminotransferase}} \text{5-phosphoribosylamine}$$

   (1) Inhibited by? → Downstream end products—IMP, GMP, AMP
7. **What inhibits de novo purine synthesis?**
   a. **Negative feedback** from naturally produced end-products
      (1) AMP and GMP both inhibit the first step in their own synthesis from IMP
      (2) IMP, AMP and GMP all inhibit upstream enzymes, specifically:
         (a) PRPP synthetase
         (b) Glutamine PRPP aminotransferase
   b. **Certain drugs** (e.g., methotrexate, 6-mercaptopurine)
   c. **Clinical correlation:** How does the drug methotrexate inhibit purine and dTMP synthesis? → It is thought to inhibit the production of THF (tetrahydrofolate) by dihydrofolate reductase.
8. **AMP and GMP can both be converted back to IMP. This allows for balanced levels of each.**
   ex) If GMP supplies decrease, then AMP converts to IMP (which is then used to replenish GMP).
9. **So now you have monophosphate nucleotides (e.g., AMP, GMP). How do you get the triphosphate versions (ATP, GTP)?**
   a. Convert monophosphates to diphosphates.
      ex) AMP → ADP (enzyme = nucleoside monophosphate kinase)
   b. Convert diphosphates to triphosphates.
      ex) ADP → ATP (enzyme = nucleoside triphosphate kinase)

D. Purine synthesis (AMP, GMP) by the salvage pathway: Take free N bases from the breakdown of RNA and DNA and add a sugar and a phosphate to them

> **Key idea:** Salvage pathway synthesis uses PRPP as the sugar and phosphate donor
> - Purine N base + PRPP → Nucleotide
> - Depending on the specific purine involved, a different enzyme is involved.

Clinical Correlation: This is a classic case of this biochemical pathway.

> **HPI:** 2 yo child presents to clinic with **self-mutilating behavior**. He has literally bitten his tongue and lips off, in addition to gnawing his fingers. On PE, the child is developmentally delayed. **Orange uric acid** crystals are noticed on the child's undergarments. The child is unable to maintain posture and has writing similar to a Huntington's patient. The child is **hyperreflexive** on PE. Lab tests confirm **hyperuricemia** and the diagnosis is made of…?
> → Lesch-Nyhan syndrome: A deficiency of an enzyme that catalyzes a purine salvage pathway reaction.
> - X-linked recessive disease → deficiency of HGPRT enzyme → uric acid accumulation → gouty arthritis, destructive behavior.

1. **Why are salvage pathways helpful?** → B/c by starting with an existing N base, it uses much less energy than de novo synthesis.

E. Degradation of purine nucleotides: Strip the ribose and the phosphate groups away, then convert the N base to uric acid

1. **Summary of purine degradation pathway:** IMP, AMP, GMP → uric acid
2. **What happens to the uric acid?** Secreted into the bloodstream.

Clinical correlations: Adenosine deaminase (ADA) deficiency, purine nucleoside phosphorylase deficiency, and **primary gout** are all diseases of the purine degradation pathway. Can you figure out which is which in these cases?

> **HPI:** 65 yo man presents with severe pain in his **big toe**. There is no history of trauma. His pain began last night and has continued unabated, without relief from analgesics. He states that he gets this pain whenever he eats a **large meal of steak and red wine**. His medical history is significant **for kidney stones**. PE reveals a fever to 38.7° C. The big toe is swollen, inflamed, and sensitive to touch. Serum tests establish a diagnosis of…?
> → Gout.
> - Elevated serum uric acid ("hyperuricemia") → birefringent needle-shaped crystals of uric acid in synovial fluid → punched out erosive lesions on x-ray with overhanging spicules → symptoms
> → **Clinical correlation: How does the drug allopurinol help relieve gout symptoms?** → By irreversibly inhibiting xanthine oxidase and thus ↓ uric acid production (xanthine oxidase catalyzes the final step: Xanthine → uric acid)
> - **FYI: Xanthine oxidase requires what as a cofactor?** Molybdenum (Mo)

> **HPI:** 1½ yo child presents with recurrent chronic viral, fungal, protozoal, and bacterial infections and with persistent diarrhea, failure to thrive, and candidiasis. Blood work shows **few detectable lymphocytes** in peripheral blood or bone marrow. **T cells are absent and an agammaglobulinemia** is diagnosed. The child is placed in a bubble with intensive antimicrobial treatment. Increased **erythrocyte dATP along with 2'-deoxyadenosine** in urine (0.01 to 0.26 mmol/mmol creatinine) confirms the diagnosis. What is it?
> → Severe combined immunodeficiency, caused by adenosine deaminase (ADA) deficiency.
> - It is thought that selective accumulation of dATP in thymocytes and peripheral blood B cells, with resultant inhibition of ribonucleotide reductase and DNA synthesis, is probably the principal mechanism of immunodeficiency in these patients.

> **HPI:** 3 yo presents with both neurologic and immunologic problems. Neurologic symptoms include **head-lag** and excessive irritability, noted when the child was 6 months. The patient is extremely **hypotonic and developmentally retarded,** with minor spastic tetraparesis. The diagnosis is suggested by the finding of undetectable—or extremely low—blood and urine **uric acid levels.** The child's medical history is also significant for recurrent otitis media. What is the diagnosis?
> → Selective T cell immunodeficiency caused by purine nucleoside phosphorylase deficiency
> - **What enzyme is deficient in selective T cell immunodeficiency?**
> - → Purine nucleoside phosphorylase

F. PYRIMIDINES: DE NOVO PYRIMIDINE (CMP, UMP, TMP) SYNTHESIS: BUILD A NUCLEOTIDE FROM SCRATCH

1. **Summary of de novo pyrimidine synthesis pathway:**

> CO_2 + glutamine → carbamoyl phosphate → dihydroorotate → UMP → UTP → CTP

Clinical correlation (Yes, more cases. Yes!): See what defect in this pathway is present.

> **HPI:** 2 yo child presents with malaise, weakness, and listlessness, progressively worse since birth. A **megaloblastic anemia** was diagnosed several months ago, but supplemental vitamin and nutrient therapy has **not been efficacious.** The child is at the 5th percentile for height and weight. A bone marrow biopsy is **not helpful.** An astute medical student decides to analyze the urine and finds characteristic crystals, which give a diagnosis of…?
> → Orotic aciduria.
> - Autosomal recessive disorder → orotidylic decarboxylase deficiency → impaired pyrimidine synthesis → impaired hematopoiesis

2. **What's the problem with de novo pathways?** → They use a lot of energy.
3. **How is thymidine (dTMP) made?**
   a. By conversion from dUMP using thymidylate synthase.

$$dUMP \xrightarrow{\text{thymidylate synthase}} dTMP$$

   b. This reaction is blocked by **the drug 5-fluorouracil** (inhibits thymidylate synthase)
4. **Key reactions of de novo pyrimidine synthesis:**
   a. **The committed step:**

$$CO_2 + \text{glutamine} + ATP \xrightarrow{\text{carbamoyl phosphate synthetase 2}} \text{carbamoyl phosphate}$$

   (1) Stimulated by? → ATP, PRPP
   (2) Inhibited by? → UTP
   b. **The rate-limiting step:**

$$\text{carbamoyl phosphate} + \text{aspartate} \xrightarrow{\text{aspartate transcarbamoylase}} \text{carbamoyl aspartate}$$

5. **What <u>inhibits</u> de novo pyrimidine synthesis?** → Negative feedback from naturally produced end-products
   a. CTP inhibits its own synthesis by inhibiting CTP synthetase
   b. UMP and CMP both inhibit an upstream enzyme—OMP decarboxylase
   c. UTP inhibits the very first step enzyme in the pathway—CAP 2
6. **What <u>stimulates</u> de novo pyrimidine synthesis?** → ATP and GTP

G. **PYRIMIDINE SYNTHESIS (CMP, UMP, TMP) BY THE SALVAGE PATHWAY: TAKE FREE N BASES FROM THE BREAKDOWN OF RNA AND DNA AND ADD A SUGAR AND A PHOSPHATE TO THEM**

> **KEY IDEA:** Pyrimidine salvage pathway synthesis uses PRPP as the sugar and phosphate donor
> - Pyrimidine N bases + PRPP → nucleotide

1. **Why are salvage pathways helpful?** → B/c by starting with an existing N base, it uses much less energy than de novo synthesis.

H. Degradation of pyrimidines: Strip the ribose and the phosphate groups away, then convert the N base to acetyl CoA or succinyl CoA (Figure 13.1)

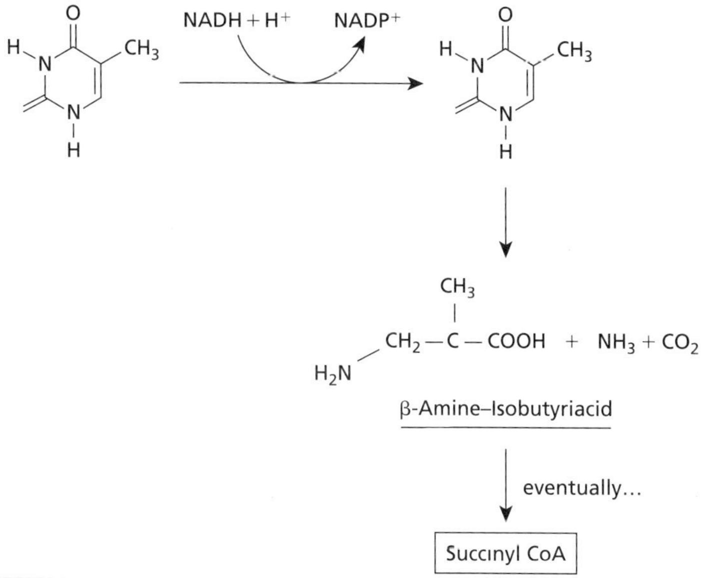

FIGURE 13.1 Degradation of pyrimidines.

CHAPTER 14
INTEGRATED REVIEW OF METABOLISM

A. **INTRACELLULAR LOCATIONS OF IMPORTANT METABOLIC PROCESSES**
   1. Cytoplasm only
      a. **Cytosol:** Glycolysis, de novo fatty acid synthesis, TG synthesis, TG breakdown, translation of proteins for intracellular use, HMP shunt, nucleotide metabolism, degradation of proteins into amino acids, some cholesterol synthesis
      b. **Endoplasmic reticulum:** Some cholesterol synthesis, translation of proteins destined for extracellular uses, steroid hormone synthesis (smooth ER), drug detoxification (smooth ER)
      c. **Golgi body:** Glycoprotein synthesis
   2. **Cytoplasm and mitochondrial matrix**

> **KEY IDEA: Three important processes occur in BOTH the cytoplasm AND the mitochondrial matrix!**
> 1. Gluconeogenesis (begins in matrix, ends in cytosol)
> 2. Urea cycle (begins in matrix, ends in cytosol)
> 3. Heme synthesis (begins in matrix, ends in cytosol)

   3. **Mitochondria only**
      a. **Intermembrane space:** H^+ gradient
      b. **Inner membrane:** Electron transport chain
      c. **Matrix:** Pyruvate decarboxylation, TCA cycle, beta oxidation of FFAs, amino acid transamination and deamination
   4. **Nucleus only:** DNA replication, mRNA transcription, RNA synthesis

B. **NAD^+ AND NADPH: TWO CRUCIAL SUBSTANCES NEEDED FOR MANY METABOLIC REACTIONS**

Nutritional correlation → What vitamin is a part of both NAD^+ and NADPH? → Niacin; niacin is the "N" in both NAD^+ and NADPH
   1. **All about NAD^+**
      a. **What is the job of NAD^+?** → To oxidize substrates in catabolic reactions

KEY IDEA: NAD$^+$ is crucial, yet your cells have only a limited supply of it.

  b. **What are the five ways that cells regenerate NAD$^+$ and where does this occur?** (Listed in order of importance.)
    (1) At the **electron transport chain** (in the mitochondrial inner membrane of all cells—except RBCs, which do not have mitochondria)
    (2) **Conversion of pyruvate to lactate** (in the cytosol of all cells):

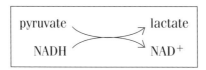

    (3) **Conversion of OAA to malate** (in the mitochondrial matrix of all cells—except RBCs, which do not have mitochondria):

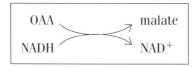

    (4) **Ketone synthesis** (in the mitochondrial matrix of liver cells only):

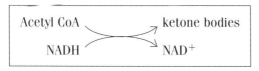

    (5) **One of the steps of TG synthesis** (cytoplasm of liver, adipose, and brush border cells):

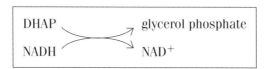

  c. **What happens when low O$_2$ conditions (e.g., exercising muscles) impair the function of the ETC?**
    (1) The #1 method for regenerating NAD$^+$ from NADH is lost!
    (2) **Dang! So, how does the cell respond?** → By increasing the use of method #2, conversion of pyruvate to lactate (This is why exercising muscles experience an increase in lactic acid concentration.)

2. **The metabolism of alcohol: An important clinical scenario involving NAD$^+$. How does the body metabolize alcohol?**
  a. Using two enzymes in the liver

KEY IDEA: The liver's metabolism of ethanol uses NAD^+ and produces NADH.

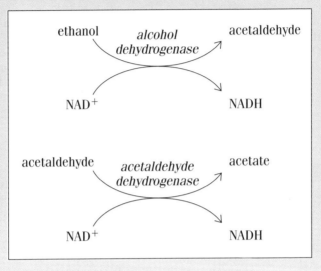

b. What is the limiting reactant in these reactions? → NAD^+

c. **Note:** <u>Alcohol dehydrogenase</u> operates with <u>zero order kinetics</u> (meaning that it metabolizes a constant number of ethanol molecules per unit time—<u>regardless of the concentration</u> of ethanol).

HPI: 23 yo woman treated with **metronidazole** for a vaginal infection presents to the ER with nausea, vomiting, and a severe headache. She has not had a fever or diarrhea. She was at a **wine tasting** recently but failed to feel a "buzz" from all the wine she was sampling. Her symptoms ensued subsequently. PE reveals **hypotension** of 95/55. All blood work is within normal limits (i.e., CBC, Chem 7, amylase, lipase).
→ Disulfiram effect (from metronidazole-inhibiting acetaldehyde dehydrogenase)
- **Clinical correlation** → **How does the drug Disulfiram (Antabuse) work?**
  → It <u>inhibits acetaldehyde dehydrogenase</u>, which leads to a buildup of acetaldehyde, which exacerbates the symptoms of a hangover. (Acetaldehyde is the molecule responsible for hangover symptoms.)
- Essentially, this drug works by making hangovers more intense and unpleasurable so that the user is driven to quit drinking. Pretty sneaky, huh?

HPI: 24 yo Korean man presents to the ER with facial flushing, increased skin temperature, headache, nausea, and vomiting. He had felt slightly strange from one half of a glass of beer at a recent medical school party. On PE, no abnormalities are found except for slight hypotension, tachycardia, and a generalized erythematous appearance. The patient is alert and oriented, and no other abnormalities are found. The attending physician fails to make the diag-

Continu

nosis. You, however, having read this book and similar case chapters earlier, make the astute diagnosis of...?
→ Aldehyde dehydrogenase deficiency
- Recent studies have shown that this flushing reaction is a result of the presence of $ALDH_2*2$, an inactive allele for the alcohol dehydrogenase gene. This leads to deficiency of the tetrameric low-Km aldehyde dehydrogenase ($ALDH_2$) enzyme that is normally found in mitochondria. Metaanalysis shows 47% to 85% of Asians versus 3% to 29% of Caucasians and other races have his genetic mutation.

3. **Clinical correlation** → Because the metabolism of ethanol <u>consumes NAD$^+$</u> and <u>generates NADH</u>, overconsumption of alcohol can cause three metabolic problems.
   a. Short answer:
      (1) Acidosis
      (2) Weight gain
      (3) Hypoglycemia
   b. Explanation: Metabolism of ethanol <u>consumes</u> NAD$^+$. Lower levels of NAD$^+$ mean two things:
      (1) ↓ Flux through reaction pathways that <u>require</u> NAD$^+$ (e.g., TCA cycle)
      (2) ↑ Flux through pathways that <u>produce</u> NAD$^+$ (e.g., conversion of pyruvate to lactate)
   c. These two changes in flux cause the following three problems via the following pathways:
      (1) **Acidosis:** ↓ [NAD$^+$] → ↓ flux through TCA → buildup of pyruvate and acetyl CoA → ↑ **formation of lactic acid and ketone bodies** (both of which are weak acids) → acidosis
      (2) **Weight gain because of** ↓ **catabolism and** ↑ **storage of fuel molecules:** ↓ [NAD$^+$] → ↓ flux through TCA → ↑ [acetyl CoA] (and) ↑ [acetyl CoA] has the following effects on certain fuel substrates (Table 14.1):

TABLE 14.1 ↑ **Acetyl CoA Causes** ↓ **Catabolism of Fuel Substrates**

Fats	Carbohydrates	Protein
• ↓ TG degradation • ↓ Catabolism of fatty acids via beta oxidation	• ↓ Flux through glycolysis • ↓ Decarboxylation of pyruvate to form acetyl CoA	• ↓ Deamination of amino acids that yield pyruvate or acetyl CoA
• ↑ De novo synthesis of FFA • ↑ TG synthesis		

   (3) **Hypoglycemia because of** ↓ **levels of gluconeogenic precursors:** Ethanol-induced hypoglycemia is a three-step process:
      (a) Metabolism of ethanol decreases the body's supply of NAD$^+$.

(b) The body responds by regenerating NAD⁺ via the conversion of pyruvate to lactate and OAA to malate:

(c) This response has one drawback—↓ [pyruvate] and ↓ [OAA] means ↓ flux through gluconeogenesis (because both pyruvate and OAA are important gluconeogenic substrates). The result can be hypoglycemia.
4. **All about NADPH**
   a. **What is the job of NADPH?** → To reduce substrates in anabolic reactions
   b. **What are the two most important means the body has for creating NADPH?**
      (1) The **hexose monophosphate shunt** (most important)
      (2) The **conversion of cytosolic malate** (this is NOT the malate in the TCA cycle, which is in the mitochondrial matrix) **to pyruvate**
   c. **What are the six crucial processes that require NADPH? (These processes are also methods for regenerating NADP⁺.)**
      (1) Steroid hormone synthesis
      (2) De novo fatty acid synthesis
      (3) Nucleotide synthesis
      (4) Regeneration of reduced glutathione (a crucial antioxidant) in RBCs
      (5) Cytochrome P450 function in the liver (which is responsible for detoxifying drugs and other foreign compounds)
      (6) Phagocytosis of foreign bodies by white blood cells (NADPH is involved in the "respiratory burst")

C. CITRATE: AN IMPORTANT METABOLIC PLAYER. THERE ARE THREE IMPORTANT METABOLIC ROLES OF CITRATE

1. **Source of acetyl CoA for de novo FFA synthesis** (via use of the citrate shuttle)
2. **Stimulates de novo FFA synthesis** by stimulating acetyl CoA carboxylase in the cytoplasm
3. **Shuts down glycolysis** by inhibiting PFK

D. PYRUVATE: A CRUCIAL SUBSTRATE AT A METABOLIC CROSSROADS

1. **The three sources of pyruvate**
   a. Glycolysis
   b. Deamination of certain amino acids
   c. Conversion from lactate
2. **The three fates of pyruvate in all cells besides the liver and kidneys**
   a. Under **low O₂ conditions: Converts to lactate** in order to regenerate NAD⁺

(1) Enzyme responsible for conversion of pyruvate to lactate? → Lactate dehydrogenase
(2) Lactate can then be sent to the liver for gluconeogenesis.
   b. Under **high O_2 conditions: Enters mitochondrial matrix and gets decarboxylated** for entry into the Krebs cycle. Enzyme responsible for pyruvate decarboxylation? → Pyruvate dehydrogenase
   c. **Converted to alanine** as a means of getting rid of intracellular NH_3
   (1) Enzyme responsible for conversion of pyruvate to alanine? → Aminotransferase
   (2) Alanine is then exported and sent to the liver, where it either enters the urea cycle or is used for gluconeogenesis.
3. **The four fates of pyruvate in a liver or kidney cell.**
   Note: The first two fates are the same as in all cells; the last two are unique.
   a. Under **low O_2 conditions: Converts to lactate** to regenerate NAD^+. Enzyme responsible for conversion of pyruvate to lactate?
      → Lactate dehydrogenase
   b. Under **high O_2 conditions: Enters mitochondrial matrix and gets decarboxylated** to yield acetyl CoA for entry into the Krebs cycle. Enzyme responsible for pyruvate decarboxylation? → Pyruvate dehydrogenase
   c. **Converted to OAA to stimulate flux through the Krebs cycle** (liver and kidney only). Enzyme responsible for converting pyruvate to OAA?
      → Pyruvate decarboxylase
   d. **Converted to OAA to enter gluconeogenesis** (liver and kidney only). Enzyme responsible for converting pyruvate to OAA? → Pyruvate decarboxylase

E. **Lactate: What you need to know**
   1. **Where does lactate originate?** → From the last step of anaerobic glycolysis—conversion of pyruvate to lactate by *lactate dehydrogenase*
   2. **Why is the creation of lactate under low O_2 conditions crucial for cell survival?**
      a. Short answer: It saves the life of the cell by regenerating NAD^+ so that glycolysis can continue!
      b. Explanation: Under low O_2 conditions, the ETC does not function at full capacity (because O_2 is required as the terminal electron acceptor). If ETC function is impaired, then reactions such as the TCA cycle and beta oxidation are NOT helpful ways to generate energy (since these reactions generate energy mainly by creating tons of NADH, which needs to go to the ETC). Without the TCA cycle and beta oxidation, the cell relies on glycolysis to generate energy. As a result, regeneration of cytoplasmic NAD^+ so that glycolysis can continue becomes absolutely crucial to cell survival!
   3. **What are the three fates of lactic acid in an exercising muscle?**
      a. Builds up within the cell
      b. Sent to the liver or kidneys to be a substrate for gluconeogenesis (this is the first part of the "Cori cycle")

c. Sent to the liver and heart muscle for entry into the TCA cycle (after first being reconverted to pyruvate)
4. **Why is a build-up of lactic acid bad?**
   a. It causes a burning sensation in exercising muscles.
   b. It is a weak acid and can thus lower blood pH (a condition called "lactic acidosis").
   c. Clinical correlation:
      (1) In what types of patients is lactic acidosis often seen?
      (2) Those with poor perfusion of tissues (e.g., from a myocardial infarction, pulmonary embolism, or hemorrhagic shock). Lactic acidosis results from the following pathway.
      (3) Poor circulation → ↓ oxygen delivery to tissues → ↑ <u>anaerobic glycolysis</u> → ↑ production of lactate

F. ACETYL CoA: ANOTHER CRUCIAL SUBSTRATE AT A METABOLIC CROSSROADS

> **KEY IDEA:** Acetyl CoA is a key metabolic intermediate. You must understand its metabolic fates and its effects on certain pathways.

1. Repeat after me: **"Acetyl CoA cannot be used in gluconeogenesis, thus fat cannot be converted into glucose."** Good. Now say it again three times. Trust me; this is important.
2. **The four sources of acetyl CoA; where does acetyl CoA originate?**
   a. **Conversion from pyruvate** (by pyruvate dehydrogenase complex)
   b. **Beta oxidation** of FFAs
   c. **Catabolism of ketone bodies**
   d. Deamination of the amino acid **isoleucine**
3. **The three fates of acetyl CoA; where does acetyl CoA go?**
   a. **Enters Krebs cycle** to generate energy (all cells)
   b. Substrate for **de novo fatty acid synthesis** (only in liver, adipose, lactating mammary, and kidney cells)
   c. Substrate for **ketone body synthesis** (which occurs only in liver cells)
4. **What metabolic effects are produced when acetyl CoA levels are <u>high</u>?**
   a. ↑ **Flux through gluconeogenesis.** Acetyl CoA stimulates the first step of gluconeogenesis (conversion of pyruvate to OAA by pyruvate carboxylase).
   b. ↑ **Flux through pathways that use acetyl CoA as a reactant.**
      (1) <u>Directly regulated pathways (i.e., acetyl CoA directly affects these reactions)</u>
         (a) ↑ Flux through **TCA cycle** thus ↑ energy generation (all cells)
         (b) ↑ **De novo synthesis of FFA** (liver, adipose, lactating mammary, and kidney cells)
         (c) ↑ **Ketone body formation** (liver cells)
      (2) <u>Indirectly regulated pathways (i.e., acetyl CoA affects these reactions via an intermediary molecule)</u>
         (a) ↑ **TG synthesis** (liver, adipose, and brush border cells)
            • Acetyl CoA indirectly stimulates TG synthesis through the following pathway: ↑ [acetyl CoA] → ↑ [FFA] → ↑ TG synthesis

c. ↓ **Flux through pathways that produce acetyl CoA as a product**
  (1) Directly regulated pathways (i.e., acetyl CoA itself affects these reactions)
     (a) ↓ **Conversion of pyruvate to acetyl CoA** (all cells)
     (b) ↓ **Beta oxidation of FFA** (all cells)
  (2) Indirectly regulated pathways (i.e., acetyl CoA affects these reactions via an intermediary molecule)
     (a) ↓ Flux through **glycolysis** (all cells): ↑ acetyl CoA → ↑ **citrate** → ↑ inhibition of PFK → ↓ glycolysis
     (b) ↓ **TG breakdown** (all cells): ↑ acetyl CoA → ↑ [FFA] → ↓ TG breakdown
     (c) ↓ **Deamination of certain amino acids:** ↑ acetyl CoA → ↑ [pyruvate] → ↓ deamination of amino acids that yield pyruvate or acetyl CoA as their carbon skeleton

5. **What metabolic effects are produced when acetyl CoA levels are low?**
→ The reverse of the previous reactions
  a. ↓ **Flux through gluconeogenesis:** ↓ acetyl CoA means ↓ stimulation of the first step of gluconeogenesis (conversion of pyruvate to OAA by pyruvate carboxylase)
  b. ↓ **Flux through pathways that use acetyl CoA as a reactant**
     (1) Directly regulated pathways (i.e., acetyl CoA itself affects these reactions)
        (a) ↓ Flux through **TCA cycle** thus ↓ energy generation (all cells)
        (b) ↓ **De novo synthesis of FFA** (liver, adipose, lactating mammary, and kidney cells)
        (c) ↓ **Ketone body formation** (liver cells)
     (2) Indirectly regulated pathways (i.e., acetyl CoA affects these reactions via an intermediary molecule)
        • ↓ **TG synthesis** (liver, adipose, and brush border cells): ↓ [acetyl CoA] → ↓ [FFA] → ↓ TG synthesis
  c. ↑ **Flux through pathways that produce acetyl CoA as a product**
     (1) Directly regulated pathways (i.e., acetyl CoA directly affects these reactions)
        (a) ↑ **Conversion of pyruvate to acetyl CoA** (all cells)
        (b) ↑ **Beta oxidation of FFA** (all cells)
     (2) Indirectly regulated pathways (i.e., acetyl CoA affects these reactions via an intermediary molecule)
        (a) ↑ Flux through **glycolysis** (all cells): ↓ Acetyl CoA → ↓ **citrate** → ↓ inhibition of PFK → ↑ glycolysis
        (b) ↑ **TG breakdown** (liver, adipose, and brush border cells): ↓ [Acetyl CoA] → ↓ [FFA] → ↑ TG breakdown
        (c) ↑ **Deamination of certain amino acids:** ↓ [Acetyl CoA] → ↓ [pyruvate] → ↑ deamination of amino acids that yield pyruvate or acetyl CoA as their carbon skeleton

G. The enzyme name game: I name the enzyme; you name the metabolic process and reaction to which it corresponds

Don't you wish someone would just make a list of all the enzymes you need to know? And then group them according to the metabolic process involved? And indicate which ones are rate limiting? Well, here you go. ☺

Note: Enzymes in **bold** are rate limiting for their respective metabolic process (Table 14.2)

Table 14.2 Summary: Most Important Enzymes in Biochemistry

Enzyme	Metabolic Process	Reaction
Glycogen synthase?	Glycogen synthesis	UDP glucose → glycogen
Glycogen phosphorylase?	Glycogen breakdown	Glycogen → glucose-1-P
Glucokinase (liver only)?	Glycolysis	Glucose → glucose-6-P
Hexokinase (nonhepatic cells)?	Glycolysis	Glucose → glucose-6-P
PFK-1?	Glycolysis	Fructose-6-P → fructose-1,6-BP
Pyruvate kinase?	Glycolysis	PEP → pyruvate
Pyruvate dehydrogenase complex?	Between glycolysis and TCA	Pyruvate → acetyl CoA
Lactate dehydrogenase?	Glycolysis under anaerobic conditions	Pyruvate → lactate
Aminotransferase?	Transamination	Pyruvate → alanine • OAA → aspartate • Alpha-KG → glutamate
Glutamate dehydrogenase?	Oxidative deamination	Glutamate → alpha-KG
Glutamine synthetase (all cells)?	Excretion of ammonia from cells	Glutamate → glutamine
Glutaminase (liver only)?	Metabolism of ammonia	Glutamine → glutamate
Citrate synthase?	TCA cycle	OAA → citrate
Isocitrate dehydrogenase?	TCA cycle	Isocitrate → alpha KG
Alpha-KG dehydrogenase?	TCA cycle	Alpha KG → succinyl CoA
Pyruvate carboxylase?	Gluconeogenesis, or stimulating flux through TCA	Pyruvate → OAA
PEPCK (phosphoenol-pyruvate carboxykinase)?	Gluconeogenesis	OAA → PEP
Fructose-1,6-bisphosphatase?	Gluconeogenesis	Fructose-1,6-BP → fructose-6-P
Glucose-6-phosphatase?	Gluconeogenesis	Glucose-6-P → glucose
Glucose-6-phosphate dehydrogenase?	HMP shunt	Glucose-6-P → 6-P-gluconolactone
Fatty acetyl CoA synthetase (thiokinase)?	TG synthesis	Fatty acid → fatty acetyl CoA
Acyltransferase?	TG synthesis	Attaches a fatty acetyl CoA to a glycerol-P

Contin

Table 14.2 Summary: Most Important Enzymes in Biochemistry—cont'd

Enzyme	Metabolic Process	Reaction
Hormone sensitive lipase?	TG breakdown	TG → fatty acetyl CoA + glycerol-P
Acetyl CoA carboxylase?	De novo FFA synthesis	Acetyl CoA → malonyl CoA
Fatty acid synthase?	De novo FFA synthesis	Acetyl CoA + malonyl CoA → palmitate
Carnitine acyltransferase 1?	Beta oxidation	Not important enough to know
Carbamoyl phosphate synthetase 1?	Urea cycle	$CO_2 + NH_4 + 2\ ATP \rightarrow$ carbamoyl phosphate
HMG CoA synthase?	Ketone synthesis	Acetoacetyl CoA → HMG CoA
HMG CoA reductase?	Cholesterol synthesis	HMG CoA → mevalonate
Desmolase?	Steroid hormone synthesis	Cholesterol → pregnenolone
Aspartate transcarbomylase?	Pyrimidine synthesis	Carbamoyl phosphate → carbamoyl aspartate
PRPP glutamyl amidotransferase?	Purine synthesis	PRPP → phosphoribosylamine
ALA synthase?	Heme synthesis	Glycine + succinyl CoA → ALA

4. INTRACELLULAR SHUTTLES: CRUCIAL FOR MOVING MATERIALS BETWEEN THE CYTOSOL AND THE MITOCHONDRIAL MATRIX

In this book, we discussed seven different shuttles and what they do. **The basic idea is twofold:**

1. These shuttles play crucial roles in metabolic processes that involve <u>both</u> the cytosol <u>and</u> the mitochondrial matrix.
2. The role they play is to <u>overcome the impermeability of the mitochondrial inner membrane</u> to certain substances (Table 14.3).

Table 14.3 Summary: Seven Biochemical "Shuttle" Systems

Name of Shuttle	Metabolic Process Involved	Metabolic Problem	Metabolic Solution (Summary)
Malate/aspartate shuttle	Any process that generates cytosolic NADH (e.g., glycolysis)	Cytosolic NADH would like to contribute to the telectron transport chain, but it cannot cross the mitochondrial inner membrane	*First:* NADH reduces DHAP to glycerol-3-P *Second:* G-3-P reacts at the inner membrane to generate FADH2 and regenerate DHAP *Third:* $FADH_2$ takes the H to the electron transport chain *Fourth:* DHAP leaves the inner membrane and enters the cytoplasm

Continued

TABLE 14.3 Summary: Seven Biochemical "Shuttle" Systems—cont'd

Name of Shuttle	Metabolic Process Involved	Metabolic Problem	Metabolic Solution (Summary)
Glycerol-3-phosphate shuttle	Any process that generates cytosolic NADH (e.g., glycolysis)	Cytosolic NADH would like to contribute to the electron transport chain, but it cannot cross the mitochondrial inner membrane	*First:* Cytoplasmic NADH reduces cytoplasmic OAA to malate *Second:* Malate reacts at the inner membrane to generate NADH *Third:* NADH takes the H to the electron transport chain
Adenine nucleotide carrier	Electron transport chain; specifically ATP synthetase	ATP synthetase requires ADP as a substrate, but ADP cannot cross the mitochondrial inner membrane	Adenine nucleotide carrier transfers one ATP out of the matrix and one ADP into it
Phosphate carrier	Electron transport chain; specifically ATP synthetase	ATP synthetase requires Pi as a substrate, but Pi cannot cross the mitochondrial inner membrane	Phosphate carrier transports Pi into the matrix
Malate shuttle	Gluconeogenesis	Enzymes are in the cytoplasm, but the starting material (OAA) is in the mitochondrial matrix and cannot get out	*First:* OAA is converted to malate *Second:* Malate exits into the cytoplasm *Third:* Malate is converted back to OAA
Citrate shuttle	De novo synthesis of fatty acids from acetyl CoA	Enzymes are in the cytoplasm, but the starting material (acetyl CoA) is in the mitochondrial matrix and cannot get out	*First:* Acetyl CoA is converted to citrate *Second:* Citrate exits into the cytoplasm *Third:* Citrate is converted back to acetyl CoA
Carnitine shuttle	Beta oxidation of fatty acids to acetyl CoA	Enzymes are in the mitochondrial matrix, but the starting material (fatty acetyl CoA) is in the cytoplasm and it cannot cross the mitochondrial inner membrane	*First:* Cytoplasmic fatty acetyl CoA reacts with carnitine to produce O-acylcarnitine *Second:* O-acylcarnitine enters the matrix and reacts to reform fatty acetyl CoA and reform carnitine *Third:* Carnitine exits into the cytoplasm

INTEGRATED REVIEW OF METABOLISM 201

I. **METABOLIC CYCLES: CRUCIAL FOR ALLOWING THE LIVER TO SUPPLY PERIPHERAL TISSUES WITH FUELS, AND FOR ALLOWING THESE TISSUES TO EXPORT NH_{4+} TO THE LIVER**
   There are three important cycles that you need to know:
   1. **Cori cycle** (should be called the glucose-lactate cycle): **Muscle cells export lactate and get glucose in return.**
      a. In the muscle
         (1) Muscle converts pyruvate to lactate (enzyme = lactate dehydrogenase).
         (2) Muscle sends lactate to liver.
      b. In the liver
         (1) Liver sends lactate through gluconeogenesis.
         (2) Liver sends fresh glucose back to the muscle.
   2. **The glucose-alanine cycle: Muscle cells export alanine (as a means of getting rid of NH_{4+}) and get glucose in return.**
      a. **In the muscle**
         (1) Muscle adds an NH_{4+} to pyruvate to create alanine (enzyme = aminotransferase).
         (2) Muscle sends alanine to the liver.
      b. **In the liver**
         (1) Liver uptakes alanine from the blood.
         (2) Liver removes an NH_{4+} from alanine to recreate pyruvate (enzyme = aminotransferase)
            (a) The NH_{4+} enters the urea cycle
            (b) Pyruvate enters gluconeogenesis
         (3) Liver sends fresh glucose (from gluconeogenesis) back to the muscle.
   3. **The glutamate/glutamine cycle: Peripheral tissues export glutaMINE (as a means of getting rid of NH_{4+}) and get glutaMATE in return**
      a. **In all tissues besides the liver and kidneys**
         (1) Oxidative deamination produces NH_{4+}.
         (2) Cell adds an NH_{4+} to gluta**mate** to create gluta**mine** (enzyme = gluta**mine** synthetase).
         (3) Cell releases glutamine into the blood; it gets absorbed by either the liver or the kidney.
      b. **In the liver**
         (1) Liver uptakes gluta**mine** from the blood.
         (2) Liver removes an NH_{4+} from gluta**mine** to recreate gluta**mate** (enzyme = glutaminase). The NH_{4+} enters the urea cycle.
         (3) Liver either sends gluta**mate** back to the peripheral tissues **or** deaminates it into alpha-KG and NH_{4+}.
            (a) The NH_{4+} enters the urea cycle.
            (b) The alpha-KG enters TCA or gluconeogenesis.
      c. **In the kidney**
         (1) Kidney uptakes gluta**mine** from the blood.
         (2) Kidney removes an NH_{4+} from gluta**mine** to recreate gluta**mate** (enzyme = glutaminase). The NH_{4+} is excreted into the urine.
         (3) Kidney either sends gluta**mate** back to the peripheral tissues **or** deaminates it into alpha-KG and NH_{4+}.
            (a) The NH_{4+} is excreted into the urine.
            (b) The alpha-KG enters TCA or gluconeogenesis.

J. **METABOLIC CONSEQUENCES OF NOT INGESTING ENOUGH CARBOHYDRATES**

If you don't get enough carbohydrates in your diet, what are the metabolic effects?
1. Glycogen stores depleted
2. ↑ Degradation of muscle proteins for use as energy
3. ↑ Breakdown of TGs; ↓ TG synthesis
4. ↑ Beta oxidation of FFAs; ↓ de novo synthesis of FFAs
5. ↑ Ketone production (to produce energy for the CNS)

K. **NUTRITIONAL CONNECTION: THERE ARE CERTAIN BIOCHEMICAL RATIONALES THAT SUPPORT THE IDEA OF A LOW-CARBOHYDRATE DIET**
  1. **↓ Carbohydrate consumption = ↑ use of fat for energy and ↓ fat storage**
     a. "↓ Carbohydrate consumption = ↑ use of fat for energy." **How?** → By ↓ acetyl CoA levels; observe the following pathway: Low blood [glucose] → ↓ flux through glycolysis → ↓ acetyl CoA → ↑ beta oxidation of free fatty acids
     b. "↓ Carbohydrate consumption = ↓ fat storage." **How?**
        (1) **By ↓ acetyl CoA levels:** Low blood [glucose] → ↓ flux through glycolysis → ↓ acetyl CoA → ↓ de novo FFA synthesis → ↓ TG synthesis and ↑ TG breakdown
        (2) **By ↓ insulin secretion:**
           (a) Low blood [glucose] → ↓ [insulin] → ↓ de novo FFA synthesis → ↓ TG synthesis and ↑ TG breakdown
           (b) Also…↓ [insulin] → ↓ uptake of glucose into adipose cells → ↓ glycolysis → ↓ [DHAP] → ↓ glycerol-phosphate → ↓ ability of adipose cells to synthesize TGs
        (3) **By ↓ levels of two important enzymes involved in de novo FFA synthesis:** Low blood glucose → ↓ acetyl CoA carboxylase and ↓ fatty acid synthase → ↓ fatty acid synthesis → ↓ TG synthesis
  2. **Low blood glucose stimulates HGH release.** Remember that low blood glucose is one of the stimuli for HGH release; and recall that two of the effects of HGH are ↑ TG breakdown and ↑ muscle protein synthesis (i.e., HGH causes you to <u>lose</u> fat and <u>gain</u> muscle).

L. **NUTRITIONAL CONNECTION: THERE ARE CERTAIN BIOCHEMICAL RATIONALES AGAINST THE IDEA OF A LOW-CARBOHYDRATE DIET**
  1. **If the body's intake of <u>carbohydrate falls too low</u>, the body's <u>ability to convert fat into energy actually becomes impaired</u>. Why?**
     a. B/c of ↓ TCA cycle activity in the liver and kidneys due to ↓ OAA levels
     b. Quick review: Remember that carbohydrates flowing through glycolysis are the body's main source of pyruvate; and pyruvate is needed to keep the TCA cycle running at full capacity in the liver and kidneys.
     c. Expanded review: The conversion of triglycerides into energy requires the TCA cycle. The TCA cycle, in turn, requires sufficient levels of pyruvate.

Pyruvate is crucial because in the liver and kidneys it forms oxaloacetate, when needed, to keep the TCA cycle running at full capacity. (Why only the liver and kidneys? → B/c these are the only organs with *pyruvate carboxylase*, which is the enzyme that converts pyruvate to OAA.)
  d. Bottom line: ↓ carbohydrates → ↓ pyruvate → ↓ oxaloacetate → ↓ TCA cycle function in liver and kidneys → ↓ conversion of fatty acids into energy

2. **As the body's <u>intake of carbohydrate falls</u>, the <u>use of protein as an energy source increases</u>.** If this increased demand for protein is not adequately met through increased dietary protein intake, the body will break down its own proteins for energy. The most noticeable result will be a loss of lean muscle mass.

Chapter 15
Nutrition

This is a very important chapter on the boards! Nutritional deficiency is one of the most clinically relevant subjects in biochemistry.

A. **High-yield concepts in nutrition**
1. **If the body's intake of <u>carbohydrates is too low</u>, the body's <u>ability to convert fat into energy is impaired</u>. Why?**
   a. Short answer: B/c carbohydrates flowing through glycolysis are the body's main source of pyruvate; and adequate levels of pyruvate are necessary to keep the Krebs cycle running at full capacity.
   b. Long answer: The conversion of triglycerides into energy requires the Krebs cycle. The Krebs cycle, in turn, requires sufficient levels of pyruvate. Pyruvate is crucial because it can form oxaloacetate, when needed, to keep the Krebs cycle running at full capacity.
   c. Bottom line: Not enough carbohydrates → not enough pyruvate → not enough oxaloacetate → ↓ Krebs cycle function → ↓ fat metabolism
2. As the body's **intake of carbohydrates falls, the use of protein as an energy source increases.** If this increased demand for protein is not adequately met through the diet, the body will break down its own proteins for energy. The most noticeable result will be a loss of lean muscle mass. Remember the Atkins diet.
3. **What vitamins and minerals are needed for proper red blood cell production?**
   a. B vitamins → B_6, B_{12}, folic acid
   b. Minerals → Copper, iron
   c. **Clinical correlation:** → Deficiencies of these substances are all possible causes of anemias.
4. **Three stages of iron deficiency—from mild-to-most severe**
   a. Depletion of iron stores...signaled by ↓ ferritin levels; ↑ transferrin levels
   b. Deficiency without anemia...signaled by low energy (b/c of changes in cytochrome function)
   c. Microcytic anemia...b/c cannot synthesize heme

5. **Common causes of B_{12} deficiency**
   a. Malabsorption
      ex) Spure, enteritis
      ex) Lack of intrinsic factor
      ex) Absent or damaged terminal ileum (which is the site of absorption)
      ex) Crohn's disease, surgical resection
   b. Vegan diet (B_{12} is found only in animal products)
   c. Lack of dietary folate (conversion of B_{12} to its active form requires dietary folate)
6. **The terminal ileum is a crucial site of absorption of the following nutrients:**
   a. Bile
      (1) **Who cares about bile reabsorption?**
      (2) Cholesterol and bilirubin are both structural components of bile. Thus bile reabsorption = cholesterol reabsorption = bilirubin reabsorption.
   b. **Folate**
   c. **Vitamin B_{12}**
   d. **Zinc**
   e. **Clinical correlation:** → If the terminal ileum is damaged or removed, a patient may develop deficiency of these nutrients.
7. A dietary pattern of <u>high intake of simple sugars</u> (with the concomitant surge of insulin) is a <u>risk factor for the development of heart disease</u>.
8. **Conversion of vitamin B_{12} to its active form requires dietary folate. Conversion of dietary folate to its active form (THF) requires B_{12}.** Therefore if levels of one are low, levels of the other automatically suffer.
9. **Because of the blood-brain barrier, <u>the brain and spinal cord have ONLY two energy sources—glucose or ketone bodies.</u>**
10. **The following links between certain diseases and foods are the only ones that have been approved by the FDA** (as of 8/99) (In official language, these are called "Health Claims.")
    a. ↓ Risk of osteoporosis and ↑ calcium intake
    b. ↓ Risk of high blood pressure and ↓ sodium intake
    c. ↓ Risk of cancer and low-fat, high-fiber, fruit-rich, and/or vegetable-rich diet
    d. ↓ Risk of heart disease and ↓ saturated fat and cholesterol intake and soluble fiber in fruits, veggies, and grains, and soluble fiber in oats and psyllium seed husk
    e. ↓ Risk of neural-tube defects and ↑ folate intake
    f. ↓ Risk of dental caries and ↓ sugar intake

B. **WHAT IS AN RDA?**
   1. RDA = "recommended daily allowance"
   2. Pay close attention to the definition: "The RDA is determined according to the judgment of the FNB (food and nutrition bureau) based on <u>available</u> data. An RDA for a given <u>essential nutrient</u> is a value that would meet, or exceed, <u>the needs</u> of <u>practically all</u> (95%) <u>healthy people</u> in a population."

This definition has several implications that point out the limitations of the RDA:
a. "Judgment of the FNB based on available data." However, the "best available data" may change as science progresses. Therefore what could be considered an RDA today may not be considered as such tomorrow.
b. "Essential nutrient." Only essential nutrients have RDAs. Many other are still important; but because they have not been defined as "essential," they do not have RDAs assigned to them.
c. "Needs." RDAs are set to avoid deficiency, not to maximize benefit. NOW, to avoid vitamin E deficiency, 200 IU may be needed; but with all we now know about the protective role of vitamin E in heart disease, 800 IU may maximize the protective benefit.
d. "Practically all (95%)." However, not all! Some people (5% of the population) will need more than the RDA.
e. "Healthy people." However, people who are sick may need much more than the RDA of a certain nutrient.

C. THE NAME GAME: COMMON NAMES AND SCIENTIFIC NAMES OF IMPORTANT VITAMINS (Table 15.1)

TABLE 15.1

Common Name	Scientific Name	Common Name	Scientific Name
Vitamin E	Tocopherol	Vitamin B_1	Thiamine
Vitamin A	Retinol, betacarotene	Vitamin B_2	Riboflavin
Vitamin D	Calciferol	Vitamin B_3	Niacin
Vitamin C	Ascorbic acid	Vitamin B_5	Pantothenate
Vitamin B6	Pyridoxine	Vitamin B_{12}	Cobalamin

D. IMPORTANT VITAMINS AND THEIR BIOLOGICALLY ACTIVE FORMS
1. Thiamin is the T in TPP (thiamine pyrophosphate).
2. RiboFlavin is the F in FADH and FMN.
3. Niacin is the N in NAD.
4. PantothenAte is the A in acetyl CoA.
5. Folate is the F in THF (tetra-hydro-folate).

E. WHAT ARE THE UNIQUE CHARACTERISTICS OF WATER-SOLUBLE VITAMINS?
1. They leech out of food, and into water, when boiled.
2. They are absorbed directly into the small intestine.
3. They are carried in blood by albumin.
4. The body has NO means of storing water-soluble vitamins (except B_{12}—which is stored in the liver).
5. Because they cannot be stored, any vitamins that are not promptly used are excreted in the urine.
6. Because they cannot be stored, frequent doses are needed to replenish what is used or excreted.

F. What are the unique characteristics of fat-soluble vitamins (A, D, E, and K)?

1. They are absorbed along with fats in the small intestine
   a. Therefore absorption is improved by concomitant consumption of fats.
   b. Therefore any mechanism of fat malabsorption (e.g., insufficient bile secretion) will also cause malabsorption of these vitamins.
2. They are stored in fat cells.
   a. Because they can be stored, they do not have to be as frequently consumed as do water-soluble vitamins.
   b. Because these vitamins accumulate in fat, toxicity is more common than with water-soluble vitamins.

G. Important nutrients: Their functions and mechanisms

Note: A few of the phrases used in the following chart require some explanation.

1. "Energy releaser" means that the vitamin plays a role in one of the processes of cellular energy generation (e.g., glycolysis, Krebs cycle, electron transport chain, B oxidation).
2. "Antioxidant" means that the vitamin reacts with free radicals and neutralizes them.
3. "Red blood cell formation" means that the vitamin plays an indispensable role in the formation of healthy red blood cells.
4. "Cell division" means that the vitamin plays an indispensable role at some point in the process of cell replication and division.
5. "Amino acid metabolism" means the vitamin has a role in one or both of the following important reactions:
   a. **Transamination:** Conversion of one amino acid to another
   b. **Deamination:** Conversion of an amino acid to a form that can be used in cellular energy generation processes
6. **Water-soluble vitamins** (Table 15.2)

Table 15.2 Water-Soluble Vitamins: Their Functions and Mechanisms

Vitamin	Function	Mechanism
Thiamin (B_1)	Energy releaser	• Essential for Krebs cycle b/c TPP is a cofactor in decarboxylation rxns (e.g., pyruvate → acetyl CoA)
Riboflavin (B_2)	Energy releaser	• Helpful for Krebs cycle and B oxidation b/c FAD is an oxidizing agent that participates in certain reactions • $FADH_2$ contributes to ATP synthesis by getting oxidized at the electron transport chain
Niacin (B_3)	Energy releaser	• Essential for glycolysis, Krebs cycle, and B oxidation b/c NAD is an oxidizing agent necessary for many reactions • NADH contributes to ATP synthesis by getting oxidized at the electron transport chain
Pantothenate (B_5)	Energy releaser	• Essential for Krebs cycle b/c it is a component of acetyl CoA and succinyl CoA (pantothenAte is the "A" in CoA)

Continued

TABLE 15.2 Water-Soluble Vitamins: Their Functions and Mechanisms—cont'd

Vitamin	Function	Mechanism
Biotin	Energy releaser	Helpful for Krebs cycle b/c it catalyzes the creation of certain substrates (e.g., oxaloacetate) by acting as a cofactor in carboxylation reactions
Pyridoxine (B_6)	• Energy releaser	• Cofactor necessary for glycogenolysis
	• Amino acid metabolism	• Cofactor for transamination and deamination reactions
	• Red blood cell formation	• Cofactor for heme synthesis
Cobalamin (B_{12})	• Cell division	• Cofactor necessary for synthesis of THF (tetra-hydro folate) from folate
	• Red blood cell formation	• Cofactor necessary for synthesis of THF from folate
	• Nervous system	• Helps with myelin sheath maintenance
Folic acid	• Cell division	• THF is necessary for synthesis of purine nucleotides and thymidine nucleotides
	• Red blood cell	• THF is necessary for synthesis of purine nucleotides and thymidine nucleotides
	• Amino acid formation metabolism	• THF is necessary for one carbon transfers
Vitamin C (ascorbic acid)	• Tissue integrity	• Necessary for collagen synthesis (vitamin C cross-link collagen)
	• Antioxidant	• Floats around and neutralizes free radicals • Participates in the regeneration of the active form of vitamin E

7. **Fat-soluble vitamins** (Table 15.3)

TABLE 15.3 Fat-Soluble Vitamins: Their Functions and Mechanisms

Vitamin	Function	Mechanism
Vitamin A (retinol) (Note: beta-carotene is a precursor)	• Vision	• As retinal, it is a constituent of visual pigments
	• Antioxidant	• Beta carotene neutralizes free radicals
Vitamin D	• Bone health	• Increases intestinal absorption of calcium by stimulating synthesis of calcium binding protein
	• Maintenance of plasma calcium levels	• Working in combination with PTH, it increases calcium reabsorption in the renal distal tubules • Working in combination with PTH, it causes bone demineralization by increasing osteoclast activity

Continued

TABLE 15.3 Fat-Soluble Vitamins: Their Functions and Mechanisms—cont'd

Vitamin	Function	Mechanism
Vitamin E	• Antioxidant	• Neutralizes free radicals
	• Heart disease	• Antioxidant function has been shown to inhibit development of atherosclerosis (proposed mechanism is protection of LDLs from free radical oxidation)
	• Red blood cell survival	• Antioxidant function also protects red blood cells from hemolysis
Vitamin K	Blood clotting	• Catalyzes an important reaction in the clotting cascade

8. **Minerals** (Table 15.4)

TABLE 15.4 Minerals: Their Functions and Mechanisms

Element	Function	Mechanism
Iron	• Energy releaser	• As cytochrome oxidase, it is part of the electron transport chain
	• Red blood cell formation	• Iron is the main part of heme (which is used to form hemoglobin and myoglobin)
Copper	• Energy releaser	• Cytochrome C oxidase is part of the electron transport chain
	• Antioxidant	• Cofactor in an antioxidant enzyme (supraoxide dismutase) that neutralizes free radicals
	• Red blood cell formation	• Necessary for heme synthesis
Zinc	• Energy releaser	• Coenzyme with lactate dehydrogenase; participates in glycolosis
	• Antioxidant	• Cofactor in an antioxidant enzyme (supraoxide dismutase) that neutralizes free radicals
	• Blood pH	• Coenzyme in certain blood buffering reactions
	• Endocrine system	• Coenzyme in insulin synthesis
	• Cell division	• Coenzyme necessary to synthesis of certain nucleotides
Calcium	• Bone health	• Structural component of bone (along with magnesium)
	• Nervous system	• Calcium influx is responsible for inducing vesicle fusion and subsequent release of neurotransmitter into synapse
	• Heart	• Calcium influx is part of myocite and pacemaker action potentials
		• Calcium release from sarcoplasmic reticulum necessary for myocite contraction

Continued

TABLE 15.4 Minerals: Their Functions and Mechanisms—cont'd

Element	Function	Mechanism
Calcium (cont'd)	• Smooth muscle	• Calcium-calmodulin complex activates MLCK, which phosphorylates myosin and causes contraction
	• Skeletal and cardiac muscles	• Complexes with troponin to expose myosin binding sites on actin filaments
	• Endocrine system	• Calcium is part of the IP3/DAG-linked second messenger system
	• Blood clotting	• Cofactor for reactions in blood clotting cascade
Magnesium	• Bone health	• Structural component of bone (along with calcium)
Chromium	• Energy releaser	• Involved in deamination reactions by which amino acids enter krebs cycle
	• Amino acid metabolism	• Involved in transamination and deamination reactions
	• Endocrine system	• Necessary for binding of insulin to cells (part of glucose tolerance binding factor)
Glutathione	• Antioxidant	• Reactant molecule necessary for regeneration of the active form of vitamin C and vitamin E
Glutathione peroxidase	• Antioxidant	• Catalyzes the regeneration of the active form of vitamin C and vitamin E using glutathione
Selenium	• Antioxidant	• Coenzyme that works with glutathione peroxidase in the regeneration of the active form of vitamin C and vitamin E using glutathione

9. **Summary: Important physiologic processes and the vitamins involved**
   a. Cellular energy generation → main nutrients involved? → All B vitamins; also, iron, copper, zinc, chromium
   b. Red blood cell formation → main nutrients involved? → B_6, B_{12}, folic acid; also, copper, iron
   c. Bone health → main nutrients involved? → Vitamin D, calcium, magnesium
   d. Antioxidants → main nutrients involved? → Vitamins A, E, C; also, zinc, copper, glutathione, selenium
   e. Blood clotting cascade → main nutrients involved? → Vitamin K (and) calcium

H. **Everyone's favorite game: Name that deficiency (as it will appear on the boards!)**

Nutritional deficiency is one of the most clinically relevant subjects in biochemistry. You will see these patients on the boards!

I. **Quick reference summary** (Table 15.5)

TABLE 15.5 Recognizing and Treating Vitamin Deficiencies

Constellation of Symptoms	Deficient Vitamin(s)	Possible Dietary Remedy
Pale color; lethargy; fatigue, shortness of breath	Iron (moderate deficiency)	Meat, fortified pasta
Night blindness; skin is whitish, flaky and peeling	Vitamin A (or betacarotene)	Spinach, fortified milk
Bending bones in children ("rickets"), bones in adults ("osteomalacia"), hypocalcemic tetany	Vitamin D	Fish, fortified dairy foods soft
Low energy, weight loss, impaired growth, muscle weakness, neurological problems	Vitamin B	Whole grains, meat, eggs
Bleeding/swollen gums, bruising, poor wound healing, anemia	Vitamin C	Citrus fruits, bell peppers
"Pins and needles" paresthesia), tetany, bone fractures, bone pain, soft bones	Calcium	Dairy
Diarrhea, dementia, dermatitis, inflamed tongue	Niacin	Whole grains, meat
Macrocytic, megaloblastic anemia (often with neurologic symptoms)	Folic acid	Green leafy vegetables, meat
Enlarged thyroid gland ("goiter"); impaired growth and mental development ("cretinism")	Iodine	Seafood, iodized salt
Impaired blood clotting	Vitamin K	Green vegetables
Neural tube defects (maternal deficiency)	Folic acid	Green leafy vegetables, meat
Mental confusion, poor muscle coordination, unsteady gait, loss of eye coordination ("Wernicke-Korsakoff syndrome")	Thiamin (moderate deficiency)	Whole grains, meat
Smooth, purple, inflamed tongue ("glossitis"); inflammation and cracking at corners of mouth ("angular stomatitis")	Riboflavin	Dairy products, eggs
Nervousness, irritability, depression	Vitamin B_6	Whole grains, nuts
Impaired growth, hypogonadism; mental lethargy; dry, scaly skin	Zinc	Meat, eggs, seafood
Microcytic anemia	Iron, copper, or B_6 severe deficiency (all are necessary for heme synthesis)	Iron—meat, green vegetables Copper—meat, seafood, nuts B_6—whole grains, nuts
Edema, congestive heart failure, muscle weakness, foot drop, wrist drop	Thiamin (severe deficiency) ("wet" beriberi)	Whole grains, meat
Muscle weakness, foot drop, sudden muscle contractions b/c of nerve damage	Thiamin (severe deficiency) ("dry" beriberi)	Whole grains, meat

J. More high-yield patients

1. **Most common vitamin deficiency in the United States?** → Folic acid
2. **Susie is an <u>alcoholic</u>. What nutrient deficiencies concern you?**
   a. Thiamin (Alcoholics have enhanced thiamin secretion.)
   b. Magnesium
3. **John is a <u>vegan</u>. What nutrient deficiency concerns you particularly?** → B_{12} (found only in animal products)
4. **Katrina has a <u>problem absorbing fats</u>. What potential nutrient deficiencies concern you?**
   a. Vitamins A, D, E, and K (absorbed along with fats in the SI)
   b. Magnesium
5. **Sally is <u>pregnant</u>. Supplementation with what vitamin is particularly essential?** → Folic acid (to prevent neural tube defects)
6. **Bill is taking <u>long-term, high-dose antibiotics</u>. This puts him at risk for what nutrient deficiencies?**
   a. Vitamin K (Other than diet, the main source of vitamin K is synthesis by bacteria in the intestines.)
   b. Biotin
7. **Chris has <u>low folate levels</u>. What other deficiency may be produced as a result?** → B_{12} (Remember: For B_{12} to be converted to its active form it needs folate!)
8. **Kim has low B_{12} levels. What other deficiency may be produced as a result?** → THF (tetrahydrofolate) (Remember: For folate to be converted to THF it needs B_{12}!)

INDEX

A

Absorptive period, 69, 78, 85-89
  versus postabsorptive period, 69
ACAT, 157, 159
Acetaminophen, 59
Acetyl CoA, 83
  aerobic glycolysis and, 124
  beta oxidation and, 143
  carbohydrate metabolism and, 131-133
  cholesterol and, 157-158
  citrate and, 82, 130, 194
  de novo FFA synthesis and, 140
  gluconeogenesis and 126, 127, 196-197
  glucose homeostasis and, 118, 119
  glycolysis and, 122, 123
  ketone synthesis and, 134
  low-carbohydrate diets and, 202
  TCA cycle and, 96, 97, 100, 101
Acetyl CoA carboxylase, 141
Acid dissociation constant, 3-4
Acid-base, 3-7
Acid-conjugate base system, 3-4
Acidic solution, 4
Acidosis, 7, 193
  lactic, 119-120, 131, 196
Aciduria, orotic, 187
Acromegaly, 95
Activated enzyme, 1
Activation energy, 54
Activator protein, 19-20
Acute intermittent porphyria, 41-42
Acute kidney failure, 177
Acyclovir, 12, 17
Acyl CoA dehydrogenase, 143
Acyltransferase, 146
Adenine, 8, 10
Adenine nucleotide carrier, 107, 108-109, 200
Adenosine deaminase deficiency, 186, 187
Adenylate cyclase, 2, 63, 66-67
Adipose, 75, 78, 81, 86, 90
Adipose cells, HMP shunt and, 183
ADP-ATP translocase, 107

Adrenal glands, 164
Adrenal hyperplasia, 165
Adrenal medulla, 73, 140
Aerobic glycolysis, 119, 120, 124
Affinity, 1
Affinity chromatography, 28
ALA dehydratase, 42
ALA synthase, 41
Alanine
  gluconeogensis and, 125
  glycolysis and, 122, 123
  pyruvate and, 82, 130, 195
  transamination and, 172
Albumin, 151
Alcohol, metabolism of, 191-193
Alcohol dehydrogenase, 61, 192
Aldehyde dehydrogenase deficiency, 192-193
Aldolase B, 115
Aldosterone, 162
Alkaline solution, 4
Alkalosis, 7
Allopurinol, 186
Allosteric enzymes, 56-58
Allosteric regulation of enzymes, 101
Allostericity, 56-58
Alpha chain, 37
Alpha helix, 33, 35
Alpha-1-antitrypsin deficiency, 33
Alpha-KG
  structure and roles of, 169
  TCA cycle and, 95, 100
  transamination and, 171-173
Alpha-KG dehydrogenase, 98-100, 101
Amino acid(s)
  carbon skeletons and, 167
  classification of, 169-170
  deamination of, 49
  degradation of, 171, 177-179
  glucagon and, 89
  homeostasis of, 177-179
  insulin and, 85
  metabolic classifications of, 170
  structure of, 31
  synthesis of, 96, 171, 178-179
  TCA cycle and, 96
  transamination and, 171-173

Amino acid transferase, 195
Aminoglycosides, 22
Aminotransferase, 82, 130
Ammonia, 93, 173-180
AMP, 112, 127, 184-185
Amytal, 103, 109, 110
Anabolic hormones, 80
Anabolic processes, 79
Anabolic reactions, 69
  versus catabolic reactions, 69
Anabolic states, 79
Anabolism, 78
Anaerobic exercise, 73-75
Anaerobic glycolysis, 119, 120
Androgens, 79, 80, 162-163, 165
Anemia
  microcytic, 42, 204
  pernicious, 54
Annealing, 27
Antabuse, 192
Antibacterial drugs, 22
Anticodon, 22
Antimycin A, 103, 109, 110
Antioxidants, 207, 210
Antivirals, 11
Apoprotein C-2, 149, 150, 160
Apoprotein CTEP, 160, 161
Apoprotein E, 150
Apoproteins, 160, 161, 162-163
Aromatase, 164, 165
Ascorbic acid, 208
Aspartate, 179, 180
Aspirin, 59, 109, 110
ATP
  allosteric enzyme regulation and, 101
  beta oxidation and, 143
  electron transport chain and, 103
  generation of, 72, 101-110
  glycogen synthesis and, 112
  glycolysis and, 122, 123
  sources of, 75
  TCA cycle and, 97
ATP synthase, 73
ATP synthetase, 102, 105
ATP synthetase inhibitors, 109, 110
Atractyloside, 109, 110
Azide, 105, 109, 110

B

Bacteria, 16, 29-30
Base pairing, 10
Beta oxidation, 69
  de novo FFA synthesis and, 145, 197
  of free fatty acids, 71
  glucose homeostasis and, 118
  inhibition of, 141
  lipid and cholesterol metabolism and, 142-145
Beta pleated sheet, 33, 36
Betacarotene, 210
Beta-hydroxyacl CoA dehydrogenase, 143
Beta-thalassemia, 58
Bicarbonate, 6
Bile, 42-43, 152-153, 158, 205
Bile salts, 152
Bilirubin
  conjugated, 43, 155
  degradation of heme and, 42-43
  detoxification of injurious compounds and, 72
  lipid and cholesterol metabolism and, 152, 153-156
  unconjugated, 43, 155
Binding, 16
  cooperative, 45, 56
Biochemistry, general principles of, 1-2
Biotechnology, techniques of, 26-30
Biotin, 141, 208
Bleomycin, 15, 17
Blood clotting cascade, 210
Blood glucose, 70, 84, 85, 89
Bone health, 210
Brain, 93-94
  glucose requirement of, 83
  glucose uptake and, 84, 116
  plasma FFAs and, 93, 140
Buffers, 4-6

C

Ca^{++}, 112
Caffeine

Caffeine—cont'd
   gluconeogenesis and, 85
   glycogen degradation and, 112
   glycolysis and, 122, 123
Calcium, 209, 211
cAMP, 2, 63-67
Cancer, 25, 205
Carbamoyl phosphate, 179-180
Carbohydrate
   anabolism and, 79
   catabolism and, 74
   deficiency of, 116-117, 202
   diets low in, 202-203, 204
   as energy source, 70
   gluconeogenesis and, 85
   glycolysis and, 85
   liver function and, 92
   metabolism of, 111-138
   TCA cycle and, 96
Carbon dioxide, 6
   affinity of hemoglobin and, 46
   hemoglobin and, 47
   TCA cycle and, 97
Carbon monoxide, 47, 105, 109, 110
Carbon monoxide poisoning, 47-48
Carbon skeletons, 70, 96, 167, 172
Carbonic anhydrase, 6
Carboxyl group, protein structure and, 31
Cardiac muscle
   absorptive period and, 78
   fuel catabolism and, 75
   lactate and, 82, 130
   myoglobin and, 44-45
   starvation and, 81
Carnitine acyltransferase, 141, 143
Carnitine shuttle, 142, 143-144, 200
Catabolic hormones, 76-78
Catabolic reactions, 69
   versus anabolic reactions, 69
Catabolic states, 76
Catabolism, 73-78
Catalyzed reaction, 55
Cataracts, 135
CDNA, 26
Cellular energy generation, 210
Cellular second messenger systems, 62-68

Central nervous system, 75, 78, 81
Ceramide, 152
cGMP, 67
Chain termination, 11
Chloramphenicol, 22
Cholesterol, 157-158
   circulation of, 158-160
   control of, 160, 161
   forms of, 157
   homeostasis of, 93
   metabolism of, 139-166
   roles of, 157
   synthesis of, 71, 157-158
   transport of, 160, 161-162, 163
Cholesterol esters, 149, 150, 157
Choroid plexus, 84
Chromatography, 28
Chromium, 210
Chylomicron remnants, 149
Chylomicrons
   apoproteins and, 163
   cholesterol transport and, 158
   function of, 161
   TG transport and, 148-149
Ciprofloxacin, 15
Cisplatin, 14, 17
Citrate
   de novo synthesis of fatty acids and, 96, 141
   glycolysis and, 122, 123
   roles of, 82, 130, 194
Citrate shuttle, 141, 200
Citrate synthase, 101
Clindamycin, 22
Cloning, DNA, 29
Cobalamine, 208
Codons, 23
Coenzyme Q, 103, 105
Coenzymes, 54
Cofactors, 53
Cold/exposure, catabolism and, 76
Collagen, 37-38
Collagen fibrils, 37
Competitive inhibition, 59, 60
Complex 1, 103, 105
Complex I deficiency, 106
Complex 2, 103, 105
Complex 3, 103, 105

Complex III deficiency, 106
Complex 4, 104, 105
Complex IV deficiency, 106
Complex 5, 105
Complimentary DNA, 26
Concentration gradients, 47
Conjugate bases, 3-4
Conjugated bilirubin, 43, 155
Constitutively active receptor, 64
Cooperative binding, 45, 56
Copper
    deficiency of, 37, 211
    electron transport chain and, 103
    function and mechanism of, 209
Cori cycle, 201
    exercise and, 133
    generation of ATP and, 71
    lactate and, 82, 120
Cori's disease, 114
Cortisol
    blood glucose and, 84, 116, 117, 118
    effects of, 77
    gluconeogenesis and, 127
    glycolysis and, 122, 123
    TG degradation and, 147-148
    TG synthesis and, 146-147
Covalent bonds, 32-33, 34
Covalent disulfide bonds, 34
Covalent peptide bonds, 34
Creatine kinase, 73
Creatine kinase reaction, 75
Crigler-Najjar syndrome, 44, 156
Cyanide
    ATP generation and, 109
    complex 4 and, 105
    as electron transport inhibitor, 110
    oxidative phosphorylation and, 110
Cyclophosphamide, 14, 17
Cystic fibrosis, 24
Cytochrome b, 103, 105
Cytochrome C, 103, 105
Cytochrome c oxidase deficiency, 106
Cytochrome c1, 103, 105
Cytochrome oxidase, 103, 105
Cytochrome p450, 61
Cytoplasm, 36, 190
Cytosine, 8, 9

Cytosine arabinoside, 12, 17
Cytosol, 190
Cytosolic malate, 194

D

Dactinomycin, 15, 17
DAG, and P3, 63, 64, 66, 67
De novo fatty acid synthesis, 83, 196
De novo FFA synthesis, 140-142
    citrate and, 82, 130, 194
    low acetyl CoA and, 197
    low-carbohydrate diets and, 202
    metabolic reactions and, 69
    versus beta oxidation, 145
De novo purine synthesis pathway, 184-185
De novo pyrimidine synthesis, 187-188
De-oxyribose, 12
Deactivated enzyme, 1
Deamination, 169
Decarboxylation of pyruvate, 72
Deep sleep, 79, 169
Denaturation, 27
Dental caries 205
Deoxyribonucleoides, 181-182
Dephosphorylation, 2
DHAP, 115
Diabetes, 118, 135-138
Diabetic ketoacidosis, 7
Didanosine, 11
Dideoxynucleotide chain termination method, 27
Dietary protein, 167-168
Dihydrotestosterone, 165
Disaccharides, 114
Disulfide bonds, 33, 34
Disulfiram effect, 192
DNA, 8-26, 29
    differences between RNA and, 12
    replication of, 15, 23
    transcription of, 23
DNA gyrase, 12, 15, 17
DNA helicase, 14, 15
DNA ligase, 15, 16
DNA looping, 19, 20
DNA polymerase, 14, 15
Down-regulation, 1

"Downstream," 1
Doxorubicin, 15, 17
DPG, 46
Dubin-Johnson syndrome, 44, 156
Duccinate dehydrogenase, 100

E

Ehlers-Danlos syndrome, 24-25, 38
Electrochemical gradient, 52
Electron transport chain(s)
    electron shuttles and, 105
    generation of ATP and, 101-110
    inhibitors of, 109, 110
    key principles of, 103-105
    NAD^+, 191
Elongation, 16, 22, 23
Elongation factor, 20-21
Emphysema, 33
Endergonic reaction, 50, 51
Endogenous circulation of cholesterol, 158-160, 162, 163
Endogenous TG transport, 148-151, 162, 163
Endoplasmic reticulum, 37, 190
Energy
    capture and storage of, 49
    creation of glucose and, 125-128
    kinetic, 54
    production of, 69
    TCA cycle and, 96, 97
Energy activation, 54
Energy intake, protein and, 168
Enhancer site, 19
Enterohepatic circulation, 152
Enteropathic circulation, 158
Enthalpy, 52-53
Entropy, 52-53
Enzyme(s), 49-61, 198-199
    allosteric, 56-58
    characteristics of, 1
    deficiencies and porphyrin pathology, 38
    inhibition of, 59-60
    mechanism of action of, 54-55
    properties of, 53-54
    rate limiting, 198-199

Enzyme(s)—cont'd
    TCA cycle and, 98
Enzyme inhibition, 59-60
Enzyme saturation, 56
Epinephrine
    blood glucose and, 117, 118
    de novo fatty acid synthesis and, 141
    effects of, 77
    gluconeogenesis and, 126-127
    glycogen degradation and, 112
    glycolysis and, 123
    ketone synthesis and, 134
    TG degradation and, 147-148
    TG synthesis and, 147
Equilibrium, 3-4
Erythromycin, 22
Erythropoietic porphyria, 38
Essential amino acids, 170
Essential fatty acids, 140
Essential fructosuria, 115
Estrogen(s), 162-163, 164-165
ETC; see Electron transport chain
Ethanol, 70, 191-193
Etoposide, 14, 17
Eukaryotic cells, 16
Exercise, 195-196
    anaerobic, 73-75
    catabolism and, 76
    lactate and, 82, 131
    metabolism and, 73
Exergonic reaction, 50, 51
Exogenous circulation of cholesterol, 158-160, 162, 163
Exogenous TG transport, 148-151, 162, 163
Exons, 13, 18
Extracellular molecules, 62-63

F

5-alpha reductase deficiency, 166
5-fluorouracil, 188
5' triphosphate group, 9
F26BP; see Fructose-2,6-bisphosphate
FAD, 100
$FADH_2$, 96, 97, 103

Fasting, catabolism and, 76
Fat homeostasis, 151
Fat-soluble vitamins, 207, 208-209
Fats
    anabolism and, 79
    catabolism and, 74
    liver fundtion and, 93
    metabolism and, 70
    TCA cycle and, 96
Fatty acids, 70, 139-140
    energy creation and, 49
    insulin and, 85
    glucagon and, 89
Fe^{2+} atom, 209
Fe^{3+} atom, 45, 48
Ferrochelatase, 42
Ferrochelatase deficiency, 38
FFA; see Free fatty acids
Fluoroquinolones, 15, 17
FMN, 103
Folate, 205
Folate deficiency, 59
Folic acid, 205, 208
Follicle-stimulating hormone, 165
Forbes disease, 114
Foscarnet, 15
Frameshift mutation, 23, 24
Free energy, 49, 53
Free energy change, 49-53
Free fatty acids, 139
    absorptive period and, 78
    beta oxidation of, 72
    glucose metabolism and, 116
    glycolysis and, 122, 123
    postabsorptive period and, 81
    synthesis of, 71
    TCA cycle and, 96
Fribroblasts, 37
Fructokinase, 114
Fructose, 114-115, 120
Fructose intolerance, hereditary, 115
Fructose-1,6-bisphosphatase, 122, 123, 126-127
Fructose-2,6-bisphosphate, 121, 123, 127, 128
Fructosuria, essential, 115
Fuel catabolism, 75

Fuel substrates, 73-79
    anabolic processes and, 79
    catabolic processes and, 74
Fumarate, 179, 180

G

G protein, 62-68
G protein–linked cell receptors, 89
G protein–linked second messenger systems, 62-67
Galactose, 114-115, 120
Gallbladder, 42
Ganciclovir, 12
Gangcyclovir, 17
Gel electrophoresis, 28
Gene, structure of, 12, 13
Genetic code, properties of, 9-12
Gigantism, 95
Gilbert's disease, 44, 156
Glucagon
    blood glucose and, 84, 117, 118
    de novo fatty acid synthesis and, 141
    effect of, on metabolism, 72, 89-92
    effect of, on target tissues, 90
    effect of insulin on, 128-129
    effects and mechanism of, 90
    gluconeogenesis and, 126, 127, 128
    glycogen degradation and, 112
    glycolysis and, 122, 123, 124
    secretion, regulation of, 91
    TG degradation and, 147-148
    TG synthesis and, 146
Glucocorticoids, 162-163
Glucogenic amino acids, 125, 169-170
Glucokinase, 119, 121, 123, 124
Gluconeogenesis, 69
    acetyl CoA and, 83, 126, 127, 196, 197
    blood glucose and, 70-71
    carbohydrate content of diet and, 85
    carbohydrate metabolism and, 125-128
    effect of insulin on, 128-129

Gluconeogenesis—cont'd
   glucose homeostasis and, 118, 119
   glucose metabolism and, 83
   intracellular locations of, 190
   need for glycogen and, 111
   pyruvate and, 82, 195
   regulation of, 127-128
   starvation and, 93
   TCA cycle and, 96
Gluconeogenic precursor, 82
Glucose, 114
   absorptive period and, 78
   anaerobic exercise and, 73
   blood, 70, 85, 89
   creation of, 125-128
   daily requirement of, 83, 94
   gluconeogenesis and, 83
   glycogen synthesis and, 112
   glycolysis and, 120
   liver function and, 92
   metabolism of, 83-85, 116-119
   starvation and, 81
Glucose homeostasis, 118-119
Glucose-6-phosphatase, 112
   deficiency of, 113
   gluconeogenesis and, 127
   glycolysis and, 120, 123
   HMP shunt and, 182
Glucose-6-phosphate
   deficiency of, 183-184
   glycogen synthesis and 112
Glucose-6-phosphate dehydrogenase, 182
Glucose-alanine cycle, 71, 72, 172, 173, 175-176, 201
Glutamate, 72, 169, 173
Glutamate/glutamine cycle, 72, 174-175, 201
Glutamine, 169, 174
Glutathione, 210
Glutathione peroxidase, 210
Glyceraldehyde, 115
Glycerol, gluconeogenesis and, 125
Glycerol phosphate, 70, 120
Glycerol-3-phosphate shuttle, 107, 200
Glycine, 35, 41

Glycogen
   degradation of, 69, 111-112
   glucose homeostasis and, 118,119
   synthesis and regulation of, 69, 71, 111-112
Glycogen storage diseases, 113-114
Glycogenolysis, 71
Glycolysis, 69
   aerobic, 119, 120, 124
   ATP and, 75
   blood glucose levels and, 71
   carbohydrate metabolism and, 70
   citrate and, 82
   generation of energy and, 49
   location and main products of, 72, 73
GMP, 184-185
Golgi body, 190
Gout, 186
Growth spurt, 79
GTP, 96, 97, 185
Guanine, 8, 10
Guanylate cyclase, 63, 66-67

H

H bonds, 33
H^+, 101-103, 105
HDL, 148-149, 150, 160, 161, 163
Heart disease, 205
Heart muscle
   absorptive period and, 78
   fuel catabolism and, 75
   lactate and, 82, 130
   myoglobin and, 44-45
   starvation and, 81
Heme
   degradation of, 42-43
   roles of, 41
   synthesis of, 39-40, 41, 93, 190
   TCA cycle and, 96
Hemoglobin, 41, 45-48
Hemoglobin-oxygen dissociation curve, 45-47
Hereditary fructose intolerance, 115
Hereditary protoporphyria, 38
Herpes family, 11, 17

Hexokinase 119, 121
Hexose monophosphate shunt, 194
HGH; *see* Human growth hormone
High blood pressure, 205
HIV therapy, 11
HMP shunt, 71, 182-184
Hormonal control of metabolism, 69, 95
Human growth hormone, 79
   blood glucose and, 84, 95, 117, 118
   glucose uptake and, 116
   metabolic effects of, 80
   protein synthesis and, 169
   synthesis of, 30
   TG synthesis of, 147
Hydrogen
   acids and, 3
   affinity of hemoglobin for oxygen and, 46
   electron transport chain and, 101-102
   DNA structure and, 9-10
   pH and, 4-5
   protein structure and, 33, 35
Hydrogen sulfide, 105, 109, 110
Hydrolysis, 52
Hydroxyurea, 181-182
Hyperammonemia, 177
Hyperglycemia, 137, 138
Hyperosmolar hyperglycemic nonketotic coma, 94
Hypertension, 205
Hypertriglyceridemia, 137
Hypoglycemia, 134, 138, 193-194
Hypoxemia, chronic, 46

IDLs; *see* Intermediate density lipoproteins
Immune system, 93
IP, 184
Infertility, 25
Inhibited enzyme, 1
Inhibition, of enzyme, 59-60
Initiation, 16, 20, 23

Insulin, 79
   cholesterol metabolism and, 141
   effects of, 72, 80, 85-89, 92
   fructose metabolism and, 114
   galactose metabolism and, 115
   glucagon and, 89, 128-129
   gluconeogenesis and, 127, 128-129
   glucose metabolism and, 116, 117
   glucose uptake and, 84, 85, 93
   glycogen synthesis and, 112
   glycolysis and, 122, 123
   high glucose levels and, 119
   low-carbohydrate diets and, 202
   mechanisms of, 86
   regulating secretion of, 87
   synthesis of, 30
   TG degradation and, 148
   TG synthesis and, 146
Insulin-dependent diabetes mellitus; *see* Diabetes mellitus
Insulinoma, 25, 94
Intermediate density lipoproteins, 150, 161, 163
Intracellular shuttles, 199-200; *see also* specific shuttle
Intracellular tyrosine kinase domain, 63
Introns, 13, 18
Iodine, 211
Ion-exchange chromatography, 28
Iron, 93, 209
Iron atom, 42, 44, 48
Iron deficiency, 204, 210, 211
Isocitrate dehydrogenase, 98, 100, 101
Isoelectric point of a protein, 32

J

Jaundice, 155-156

K

Ka value, 3-4
Kartagener's syndrome, 25
Ketoacidosis, 7, 135, 137
Ketogenic amino acids, 169-170

Ketone bodies
   glucagon and, 89
   glucose homeostasis and, 118, 119
   insulin and, 85
   starvation and, 81
   synthesis of, 83, 191, 196, 197
   types of, 133
Ketone metabolism, 133-135
Ketonemia, 135
Ketonuria, 135
Kidney(s)
   ammonia and, 176
   effect of insulin on, 86
   glucagon and, 90
   gluconeogenesis and, 125-128
   glutamate/glutamine cycle and, 174-175
   HMP shunt and, 183
   metabolic cycles and, 201
   oxidative deamination and, 173
   TCA cycle and, 100
Kidney failure, 177
Kinase, 2
Kinetic energy, 54
Km, 56, 60
Krebs cycle
   acetyl CoA and, 83, 97, 196
   generation of energy and, 49
   low-carbohydrate diets and, 204
   pyruvate and, 82, 130, 131
Kwashiorkor, 168

L

Lactate
   build up of, 82, 196
   carbohydrate metabolism and, 131
   creation of, 82, 195
   exercising muscle and, 82, 195-196
   gluconeogenesis and, 125
   glycolysis and, 120
   metabolism of ethanol and, 193-194
   NAD^+ and, 191
   pyruvate and, 82, 130, 194-195
   transfer of, 130

Lactate dehydrogenase, 82, 195
Lactic acid, 195, 196
Lactic acidosis, 119-120, 131, 196
Lactose, 114
Lagging strand, 16
Lamivudine, 11
L-B plot; see Lineweaver-Burke plot
LCAT, 157 160
LDL, 158, 161, 163
Lead poisoning, 42
Leading strand, 16
Lecithin, 152
Lesch-Nyhan syndrome, 186
Leucine, 170
Leydig cells, 164
Lincomycin, 22
Lineweaver-Burke plot, 56, 57
Linoleic acid, 140
Linolenic acid, 140
Lipid and cholesterol metabolism, 139-166
Lipids, 139
Lipogenesis, 146-147
Lipoic acid, 98, 124
Lipolysis, 147-148
Lipoprotein lipase, 147
Lipoproteins, 148, 160, 161-163
Liver
   absorptive period and, 78
   ammonia and, 176
   catabolism and, 75
   Cori cycle and, 133
   effect of insulin on, 86
   functions of, 92-93
   glucagon and, 90
   gluconeogenesis and, 125-128
   glucose homeostasis and, 118, 119
   glucose uptake and, 84, 116
   glutamate/glutamine cycle and, 174-175
   glycogen degradation and, 112
   glycogen synthesis and, 111, 112
   heme degradation and, 42
   HMP shunt and, 183
   lactate and, 82, 130
   metabolic cycles and, 201
   oxidative deamination and, 173

Liver—cont'd
  postabsorptive period and, 81
  pyruvate in, 82
  starvation and, 81, 93
  TCA cycle and, 100
  TG storage and synthesis and, 147
Low-carbohydrate diets, 202-203
Lysine, 170
Lysosomes, glycogen degradation and, 112
Lysyl oxidase, 37

M

Magnesium, 85, 210
Malate, 191, 194
Malate dehydrogenase, 100
Malate shuttle, 126, 200
Malate/aspartate shuttle, 107, 108, 199
Malonate, 103, 109, 110
Malonyl CoA, 140-141, 143
Maltose, 114
Mammary glands, 183
Marasmus, 168
McArdle's syndrome, 114
McCune-Albright syndrome, 68
Messenger RNA, 16, 18, 19, 23
Metabolic alkalosis, 7
Metabolic cycles, 201; *see also* specific cycle
Metabolism
  of alcohol, 191-193
  of ammonia, 176-177
  of carbohydrates, 111-138
  catabolic processes of, 49
  of cholesterol, 139-166
  exercise and, 73
  fats and, 70
  of fructose, 114
  of galactose, 115
  of glucose, 83-85, 116-119
  goals of, 71-72
  hormonal control of, 69-95
  ketone, 133-135
  of lipids, 139-166
  of monosaccharides, 114-115

Metabolism—cont'd
  nitrogen and protein, 167-188
  nucleotide, 181-189
  overview of, 69-95
  review of, 190-203
Methemoglobin, 48
Metmyoglobin, 45, 48
Michaelis-Menten equation, 55-56
Microcytic anemia, 42, 204
Mineralocorticoids, 162-163
Minerals
  functions and mechanisms of, 209-210
  liver function and, 93
Missense mutation, 22, 24
Mitochrondria, 41, 190
Mitochondrial inner membrane, 107
Mitochondrial matrix, 130, 190
Molecular biology, techniques of, 26-30
Monomer, 37
Monosaccharides, metabolism of, 114-115
mRNA; *see* Messenger RNA
Muscle
  ammonia and, 176
  Cori cycle and, 133
  effect of glucagon on, 90
  effect of insulin on, 86
  exercising, 73, 131, 195-196
  glucose/alanine cycle and, 175-176
  glycogen degradation and, 112
  skeletal, 84, 111, 112, 168
Mutations, 22-24
Myoglobin, 41, 44-45, 48

N

N bases; *see* Nitrogenous bases
NAD^+, 73, 82, 100, 190-194
NADH, 96, 97, 101, 103
NADH dehydrogenase, 103, 105
NADH dehydrogenase deficiency, 106
NADPH, 140, 182-184, 190-194
Neural-tube defects, 205
Neuropathy, peripheral, 135
Neutral solution, 4

NH_3, 167, 172-173
NH_4^+; see Ammonia
Niacin, 98, 124, 190, 207, 211
Niemann-Pick disease, 152
Nitric oxide, 63-67
Nitric oxide synthase, 62
Nitrogen and protein metabolism, 167-188
Nitrogenous bases, 8, 9, 181-182
Nitrosureas, 14, 17
Noncompetitive inhibition, 59, 60, 61
Nonessential amino acids, 170
Non–insulin-dependent diabetes mellitus; see Diabetes mellitus
Non-hepatic cells, pyruvate and, 82
Nonreactive substrate analogs, 59
Nonsense mutation, 23, 24,
Norepinephrine
    blood glucose and, 84, 117, 118
    glycolysis and, 123
    ketone synthesis and, 134
    metabolic effects of, 77
    TG degradation and, 147, 148
    TG synthesis and, 147
Norfloxacin, 15
Northern blot, 28
NRTIs; see Nucleoside reverse transcriptase inhibitors
Nuclear magnetic resonance, 28
Nucleoside, 181-182
Nucleoside reverse transcriptase inhibitors, 11
Nucleotide
    anabolism and, 79
    frameshift mutation and, 23
    metabolism and, 70
    purine and, 186-187
    as structural unit of DNA and RNA, 8-11
    synthesis of, 182-184
Nutrition, 204-212
Nutritional deficiencies, 210-211

O

O-acylcarnitine, 142
OAA; see Oxaloacetate

Obesity, 89
Ofloxacin, 15
Okazaki fragments, 16
Oligomycin, 105, 110
Ornithine, 179
Orotic aciduria, 187
Osteoporosis, 205
Oxaloacetate
    gluconeogenesis and, 125
    NAD^+ and, 191, 194
    pyruvate and, 82, 131, 195
    TCA cycle and, 96, 97, 100
Oxidation, 2
    beta; see Beta oxidation
Oxidative deamination, 172, 173, 177-179
Oxidative phosphorylation, 72, 101-103
Oxygen, 103-105
    lack of, 105, 110
    transport of in blood, 45-48
Oxygen dissociation curve, 45-47

P

P3 and DAG, 67
Palmitate, 140, 141
Palmitoyl CoA, 141
Panacinar emphysema, 33
Pancreas, 86, 91
Pantothenate, 98, 124, 207
PCR; see Polymerase chain reaction
PEP; see Phosphoenolpyruvate
PEPCK; see Phosphoenol-pyruvate carboxykinase
Peptide bond, 32
Peptide linkage, 32, 175
Peptidyl transferase, 22
Peripheral neuropathy, 135
Pernicious anemia, 35, 54
PFK-1; see Phosphofructokinase 1
pH, buffers and, 4-6
Phosphatase, 2
Phosphate, 85
Phosphate carrier, 107, 109, 200
Phosphatidic acid, 152
Phosphatidylcholine, 152

Phosphodiester bonds, 8, 9, 11
Phosphodiesterase, 2
Phosphoenolpyruvate, 122
Phosphoenolpyruvate carboxykinase, 126
Phosphofructokinase 1, 119, 121-122, 123, 124
Phosphoglycerides, 151-152
Phospholipase, 66-67
Phospholipase C, 63
Phospholipids, 139, 151-152
Phosphorylation, 2
Photosensitivity, 25
Physiologic buffer system, 6
Piercidin A, 103, 109, 110
Pituitary adenoma, 95
pKa, 4, 5-6
Plasmids, 16
Point mutations, 22, 24
Poisoning, lead, 42
Poly adenylate tail, 17-18
Poly guanine cap, 17-18
Polydipsia, 135
Polymerase chain reaction, 27, 28
Polymorphism, 26
Polypeptide chains, 32
Polypeptide heme chain, 42-43, 153-154
Polyuria, 135
Pompe's disease, 113
Porphyrias, 41-42
Porphyrins, 38-40, 41
　related enzyme deficiencies and, 38
Postabsorptive period, 69, 79, 81
　versus absorptive period, 69
Posttranscriptional modifications, 17-18
Precipitation, selective, 28
Progestins, 162-163
Proline, 35
Propionyl CoA, 125
Protein(s)
　activator, 19-20
　anabolism and, 79
　catabolism and, 74
　collagen and, 37-38

Protein(s)—cont'd
　degradation of, 169
　as energy source, 70
　heme and, 41-43
　hemoglobin and, 45-48
　liver function and, 92
　metabolism and, 70
　myoglobin and, 44-45
　and nitrogen metabolism, 167-188
　packaging and distribution of, 36-37
　porphyrias and, 38-40
　procedures involving, 28-30
　purification and separation of, 28
　structure of, 22-24, 28, 31-36
　synthesis of, 22, 71, 169
Protein kinase, 122
Protein kinase A, 64, 67, 123
Protein kinase C, 64, 67
Protein kinase G, 64, 67
Protein-DNA, detection of, 29
Proton gradient, 102
Protoporphyria, hereditary, 38
Proximal convoluted tubule, 84
Pseudohypoparathyroidism, 63
Purine(s), 8, 10, 168
　de novo synthesis of, 184-185
　degradation of, 186
　synthesis of, 71
Purine degradation pathway, 186-187
Purine nucleoside phosphorylase deficiency, 186, 187
Purine nucleotides, degradation of, 186-187
Pyridoxine, 208
Pyrimidine(s), 8, 9, 168
　de novo synthesis of, 187-188
　degradation of, 189
　synthesis of, 71, 188
Pyrimidine salvage pathway synthesis, 188
Pyruvate
　acetyl CoA and, 197
　aerobic glycolysis and, 124
　carbohydrate metabolism and, 130-131
　conversion of, 121

Pyruvate—cont'd
   decarboxylation of, 72
   gluconeogenesis and, 125, 195
   glycolysis and, 120
   in liver cells, 82
   low-carbohydrate diets and, 203
   NAD^+ and 191, 194-195
   in nonhepatic cells, 82
   TCA cycle and, 96, 97
   transamination and, 172
Pyruvate carboxylase, 100, 126
Pyruvate decarboxylase, 82, 195
Pyruvate dehydrogenase, 82, 98, 100, 101, 195
Pyruvate dehydrogenase complex, 124
Pyruvate dehydrogenase deficiency, 98
Pyruvate kinase, 119, 121, 122, 123, 124

Q

Quaternary structure of proteins, 33-34

R

R form of hemoglobin, 45
Rate-limiting step of reaction pathway, 1
RDA; see Recommended daily allowance
Reaction, increasing the rate of, 53-58
Reaction pathway, rate-limiting step of, 1
Reaction velocity, 1
Recombinant DNA, 29
Recommended daily allowance, 205-206
Red blood cells
   bilirubin and, 153
   glucose uptake and, 84, 116
   HMP shunt and, 183
   nutrition and, 204, 210
   plasma FFAs and, 73, 140
Reduction, 2
Replication, 8-25, 27
   key ideas of, 12

Repressor protein, 19
Restriction endonucleases, 26
Restriction enzymes, 26
Restriction fragment length polymorphism, 26
Restriction fragments, 26
Retinol, 208
Reverse transcriptase, 26
RFLP; see Restriction fragment length polymorphism
Riboflavin, 124, 207, 211
Ribonucleotides, 181
Ribose, 12
Ribose-5-phosphate, 182-184
Ribosomal RNA, 18, 19
Ribosomes, 18
Rifampin, 17
RNA, 8-25, 181
   differences between DNA and, 12
   replication of, 23
   transcription of, 23
RNA primer, 13, 16
Rotenone, 103, 109, 110
Rotors syndrome, 156
rRNA; see Ribosomal RNA

S

17-alpha-hydroxylase deficiency, 165
Salvage pathway, 186, 188
Scurvey, 38
Seabright-Bantam syndrome, 63
Second messenger systems, 62-68, 88, 89
Secondary protein structures, 33, 34, 35-36
Selective precipitation, 28
Selenium, 210
Sertoli cells, 164
Shuttle systems, 107-109, 199-200; see also specific shuttle
Silent mutation, 22, 24
Skeletal muscle
   absorptive period and, 78
   fuel catabolism and, 75
   glucose uptake and, 84, 116
   glycogen synthesis and, 111, 112

Skeletal muscle—cont'd
　myoglobin and, 44-45
　starvation and, 81, 168
　transfer of lactate from, 130
Skin cancer, 25
Sleep, deep, 79, 169
Small intestine, 42-43, 84
Southern blot, 28
Southwestern blot, 29
Sphingolipids, 151-152
Sphingomyelinase, 152
Spleen, 42
Starvation
　compared to diabetes, 137
　dietary protein and, 168
　insulin and, 89
　　ketone synthesis and, 134
　　metabolic processes during, 81
　　production of fuel substrates in, 93
Stavudine, 11
Stercobilin, 43-44, 154
Steroid hormones
　as activator proteins, 19
　synthesis of, 71, 159, 162-166
Sterols, 139
Stimulated enzyme, 1
Stress, catabolism and, 76
Strong base, 3, 4
Substrate level phosphorylation, 72, 73
Succinate dehydrogenase, 101
Succinate thiokinase, 100
Succinyl CoA, 41, 143, 145
Sucrose, 114
Sugar, 8, 205
Sulfonamides, 59

T

2,4-dinitrophenol, 109, 110
21-hydroxylase deficiency, 165
T form of hemoglobin, 45
TCA cycle, 71
　acetyl CoA and, 197
　electron transport chain and, 96-110
　gluconeogenesis and 96

TCA cycle—cont'd
　lactate and, 120
　low carbohydrate diet and, 202-203
　protein metabolism and, 70
　urea cycle and, 180
Temperature
　affinity of hemoglobin for oxygen and, 46
　effect of on free energy change, 51, 52-53
　increase in reaction rate and, 58
Termination, 16, 22, 23
Testosterone, 164-165
Tetracyclines, 22
Theca cells, 164
Thermodynamic coupling, 51
Thiamin, 98, 124, 207
Thiamin deficiency, 184, 211
Thiokinase, 146
Thiophorase, 134
Thymidine, 12, 16, 188
Thymine, 8, 9
Thyroid hormone, 77-78, 147-148
Tissue plasminogen activator, 30
Titration curve, 5, 6
Topoisomerase 2, 14, 17
TPA; *see* Tissue plasminogen activator
Transaldolase, 182
Transamination, 169, 171-173, 177-179
Transcription, 8-25
　key ideas of, 13
Transfer RNA, 18, 19
Transition state, 54
Transketolase, 182
Translation, 8-25, 32, 36-37
Triglyceride(s), 69, 139
　degradation of, 85, 89, 147-148, 197
　key ideas of, 13
　storage of, 146, 147
　synthesis of, 71, 146-147, 191, 196, 197
　transport of , 148-151, 160, 161-162, 163
tRNA; *see* Transfer RNA
Tryptophan, 35

Tylenol, 59
Tyrosine kinase activity, 66
Tyrosine kinase receptors, 88

U

Ubiquinone, 103, 105
Ubiquinone-cytochrome c
    oxidoreductase deficiency, 106
Unconjugated bilirubin, 43, 155
Uncoupling agents, 109, 110
Up-regulation, 1
"Upstream," 1
Uracil, 8, 9, 12, 16
Urea, 179-180
Urea cycle, 71, 179-180, 190
Urobilin, 43
Urobilinogen, 43, 154
Uroporphyrinogen deficiency, 38

V

Viruses, 16
Vitamin B deficiencies, 98, 100, 124
Vitamin B_{12} deficiency, 35, 54, 205
Vitamin C, 37, 38
Vitamin D, 159
Vitamin deficiencies, 211
Vitamins, 206; *see also* specific vitamin

Vitamins—cont'd
    fat-soluble, 207, 208-209
    liver function and, 93
    water-soluble, 206, 207-208
VLDL, 149-150, 161, 163
VLDL remnants, 150
Von Giercke's disease, 113, 127

W

Water-soluble vitamins, 206, 207-208
Weak base, 3, 4
Weight gain, 193
Well-fed state, 85-89
Wernicke-Korsakoff syndrome, 184
Western blot, 29
White blood cells, HMP shunt and, 183

X

Xeroderma pigmentosum, 25
X-ray crystallography, 28

Z

Zalcitabine, 11
Zidovudine, 11
Zinc, 205, 209, 211
Zona reticularis, 164
Zwitterion, 31